Frederick W. Pavy

The Croonian Lectures on Certain Points Connected with Diabetes

delivered at the Royal College of Physicians

Frederick W. Pavy

The Croonian Lectures on Certain Points Connected with Diabetes
delivered at the Royal College of Physicians

ISBN/EAN: 9783337272432

Printed in Europe, USA, Canada, Australia, Japan

Cover: Foto ©berggeist007 / pixelio.de

More available books at **www.hansebooks.com**

THE

CROONIAN LECTURES

ON CERTAIN POINTS

CONNECTED WITH

DIABETES

DELIVERED AT THE

ROYAL COLLEGE OF PHYSICIANS

BY

F. W. PAVY, M.D., F.R.S.,

FELLOW OF THE COLLEGE ;

PHYSICIAN TO, AND LECTURER ON THE PRACTICE OF MEDICINE AT. GUY'S HOSPITAL.

LONDON

J. & A. CHURCHILL, NEW BURLINGTON STREET

1878

TABLE OF CONTENTS.

CROONIAN LECTURES

POINTS CONNECTED WITH DIABETES.

Mr. President and Gentlemen,—The first words I desire to utter in these lectures are words expressive of appreciation of the merits of the great physiologist whom science has just lost. Although upon a certain point I have been forced to differ from Bernard's views, and it is upon this point that I shall have to engage your attention in the ensuing lectures, yet I cordially acknowledge the immense gain to physiological knowledge which has been acquired through his labours, and, as an old pupil, testify to his high qualities as a man of science. No one can deny that the earnest desire which has been so widely evinced to honour his memory has been most thoroughly merited. I feel that the moment is an inopportune one for disputing the validity of work that he has done, but it is beyond my control to alter the position in which I am placed. The exigencies of science must stand

1

before all personal considerations. Upon this
ground, and with the assertion, which can be truly
made, that no spirit of personal hostility prompts
me, I claim justification for the criticism I shall
have to offer, and without further words will proceed to the immediate subject of these lectures.

From time immemorial diabetes has been one of
the most inscrutable of diseases. All sorts of vague
notions have existed regarding the nature of the
affection, and at the present day it must be said that
opinions are by no means settled upon the fundamental points to be dealt with. As a close worker
for the last twenty-five years, as my time has
permitted, upon the subject, I have endeavoured by
investigation to ascertain the true state existing,
and I propose here to set forth the facts and argu-
ments which appear to me to lead up to and justify
certain definite conclusions.

First, let me give a general expression of the
outward evidence manifested to us on viewing the
diabetic as compared with the natural state.

In the healthy person, when starch and sugar—
principles which form important elements of our
food—are ingested, they become lost sight of in the
system. We know that the starch is first converted
into sugar, so that the two are brought to the same
position before absorption occurs. The sugar thus
entering the system is subsequently so acted upon
as to be rendered susceptible of being applied to the
requirements of the economy. The healthy system,
it may be said, as the result of observation, possesses a power of assimilating and subsequently
utilising the absorbed saccharine matter, and hence

we do not find it escaping in an unconsumed state from the body. In diabetes, on the other hand, there exists a want of assimilative power over the saccharine principle. From this defect, the starch and sugar of the food escape under the form of sugar, or in an unconsumed state, with the urine. Observation shows that in proportion to the starch and sugar ingested sugar is eliminated by the kidney, besides a certain amount taking its origin in another way. We have here a simple statement of fact without the introduction of any theory. In the one case the sugar disappears from view, and doubtless, as a final issue, contributes to force-production; in the other it fails to be so utilised, but passes through the system and escapes in an unconsumed state. This constitutes an essential feature of the disease, and one great object of investigation is to find out, by following the internal changes, the explanation of this result.

I believe it is only through physiology that we can expect to be able successfully to approach pathology with reference to this subject, and we must be quite sure of our ground, step by step, as we advance. We must have correct information of the natural state before we can know where we are in relation to the unnatural. As the urine gives us the outward manifestation of the effects of the disease, let us commence our˚ investigation by ascertaining what is to be learnt regarding this secretion in relation to sugar.

Of the tests for sugar, the copper solution (Fehling's, or a modification of it, the potassic-tartrate of copper, which I am in the habit of employing

myself) is the most delicate and reliable. Now, on
testing a sample of urine with this liquid, the reaction may be such as to leave no doubt that sugar
is present to a more or less notable extent. With
another sample of urine there may be no neat and
decided reaction perceptible, but a certain amount
of change suggestive of the indication of a slight
amount of sugar. Be it remarked, however, that this
behaviour may be due to the action of lithic acid,
especially should the urine be loaded with lithates, for
lithic acid is an agent which has some reducing
power over the oxide of copper. In another case,
a sample of urine may be tested, and no sign of
reaction shall be visible.

What is the true condition of these latter samples
of urine? Are we to take the doubtful and negative behaviours for what they appear to represent, or
may sugar be present, but fail to be revealed, or
distinctly revealed, by our reagent? It is important, not, it is true, with reference to any
clinical bearing, but as a physiological, and thence
pathological, consideration, that we should know
whether healthy urine is really free from sugar or
not, and I will proceed to show what evidence is
adducible upon this point.

It is known that our test agent which has been
referred to does not act with the same sensitiveness
in the presence of urinary matter as with a pure
solution of sugar. Seegen has fully pointed this
out. By this authority it has been shown that a
certain minute quantity of sugar, which, when dissolved in urine, may fail to be susceptible of being
satisfactorily revealed, may, when dissolved in water,

be seen to give a distinct reaction. Hence we have here proof that a negative behaviour of our test is not to be taken as evidence of an entire absence of sugar.

Brücke some years back proposed a process for the separation of sugar by throwing it down in combination with oxide of lead. The process is an exceedingly satisfactory one, and by its means the sugar, to however minute an extent it may be present, can be abstracted, and thus, whatever exists in a large quantity of urine, may be obtained in a separated and concentrated form. It is just the process that is wanted for showing the condition of ordinary urine in relation to sugar; and not only may it be used for qualitative, but likewise for quantitative purposes.

A few words will suffice to describe the mode of application, and in the course of my description I will illustrate the steps of the operation. The sample of urine, in any quantity that it may be considered advisable to take, is first treated with an excess of the neutral acetate of lead. The effect of this is to produce a copious precipitate, the uric acid, sulphuric acid, phosphoric acid, chlorine, and doubtless some other constituents of the urine, being carried down. The sugar remains in solution untouched. Filtration is performed, and the filtrate, which is the product with which we are concerned, is treated with ammonia and a further quantity of acetate of lead, unless a large excess of this agent has been in the first place used. The sugar now falls amongst the copious precipitate again produced. In an acid solution—and such the urine as

first taken was—the plumbic acetate does not lead
to the precipitation of sugar, but in the presence of
free ammonia a definite insoluble compound is
formed consisting of two atoms of sugar and three
atoms of oxide of lead. The object now is to collect
and wash the precipitate, and then liberate the
sugar. It is quite requisite, in view of the subsequent application of the copper test, that all the
ammonia should be removed, and the washing is
best effected by a few repetitions of subsidence and
decantation before throwing on the filter. The
washing is then carried on till the water which
passes through no longer gives a blue colour to
reddened litmus-paper. The necessity of removing
all the ammonia has been mentioned, for unless this
is thoroughly done, its presence will interfere with
the proper precipitation of the suboxide should the
copper test be subsequently employed.

Washing with hot water expedites the process by
causing the precipitate to assume a more dense form,
and thus more rapidly to subside in the process by
decantation, and also afterwards to be more readily
percolated on the filter. Experience has taught me,
however, that danger is incurred of losing some of
the sugar. The combination between the sugar and
plumbic oxide seems not to be a strong one, and
with boiling water I have reason to think that the
sugar may be entirely removed.

The washed precipitate is next treated in such a
manner as to liberate the sugar. This may be done
by the agency of hydrochloric, sulphuric, or oxalic
acid, but a somewhat coloured product is the result.
It might be suggested also that the acid may lead to

the formation of sugar, or of a substance that, like it, exerts a reducing action on the copper test, from some other constituent of the urine, whether colouring matter or not, carried down with the lead precipitate. Schunk has pointed out, and I can confirm his statement, that boiling hydrochloric acid certainly appears to act in this way. No exception, however, can be taken to the use of sulphuretted hydrogen; and with this agent a better, because a purer and more colourless, product is obtained. The precipitate with a little water is placed in a suitable apparatus and a stream of sulphuretted hydrogen passed through till decomposition is thoroughly effected, which may be known by the uniform production of intense black sulphide. With a moderate amount of precipitate a few hours will suffice for the purpose. Filtration is next performed, and the excess of sulphuretted hydrogen expelled by heat. The liquid is then brought down to a small bulk, either over the water bath or in the vacuum of an air-pump.

I have here mentioned the main points requiring to be referred to. Full details upon the whole subject I gave in a communication " On the Recognition of Sugar in Healthy Urine," inserted in the ' Guy's Hospital Reports' for 1876.

If we take the product that has been obtained, and test it with the copper test a neat reaction is produced [result shown]. Such is the result that we notice, but the question arises—Is this result to be taken as conclusive evidence that sugar is present? There can be no fallacy arising through the presence of lithic acid, which exerts a certain degree of reducing effect with the copper solution, for this principle was eliminated with the first lead precipitate in the process employed. We need not, however, rest with the action of one test only. Moore's or the liquor potassæ, and Böttger's or the bismuth test, equally give a good reaction [result shown]. Further, there is another test, which is accepted as affording crucial evidence regarding the presence of sugar—I mean fermentation, and this may be shown to give a positive behaviour. In my early efforts I failed to obtain a reaction with this test, even although I had manipulated with urine to which I had purposely added sugar. I afterwards found the failure of reaction arose from the acidity of the product yielded by the process of preparation employed. It did not occur to me at first to look to this point, but Pasteur has shown that acidity opposes, whilst alkalinity favours, the occurrence of fermentation. After removing the acidity by neutralising with carbonate of soda, I found that fermentation actively proceeded. Some of the product obtained from healthy urine has been set going with washed yeast in this fermentation apparatus which I have devised for the application of the test. It may be seen that a considerable amount of gas has been generated and is accumulated in the upper tube. I propose to show, by the action of potash, that this gas consists of carbonic acid, and the chromic acid test will reveal the presence of alcohol in a few distilled drops from the fermented contents of the lower tube [experiment shown]. A counterpart application of the test with yeast and water was started at the same time as the other, and the result shows that only a comparatively insignificant amount of gas has been generated.

The evidence, then, that has been presented ocularly before you, may be taken, I consider, as affording a demonstration of the existence of sugar in healthy urine. It is confirmatory of the results obtained several years back by Brücke and Bence Jones. We are not correct, therefore, in inferring from a negative reaction under the ordinary mode of testing, that proof has been afforded of an absence of sugar. It is true that, on examining urine and finding no reaction given by the copper test, it is customary to speak of it as free from sugar. As far as the requirements of clinical observation are concerned, such language may continue to be used, for the amount of sugar which is beyond the reach of ordinary testing in a decided manner to reveal has no clinical significance belonging to it. This I need hardly say should be taken as thoroughly understood. When precision, however, has to be considered, we are not justified in so expressing ourselves.

Not only can it thus be said that sugar exists in healthy urine, but the amount present is susceptible of being expressed in definite terms. Having satisfied ourselves by the corroborative evidence of the fermentation and other tests that we have really sugar to deal with, the information afforded by the copper solution may be accepted for quantitative purposes. There is no difficulty in obtaining a satisfactory result with a moderate quantity of urine. It is only lately that I have realised this to be the case. Even 100 cc. (about four fluid ounces) will suffice for the analysis. With double this quantity it is even easy to obtain a concentrated product, which gives a very perceptible reaction on being tested in the ordinary manner with the copper solution.

For the quantitative analysis the lead process which has been described is employed to furnish the product for making the determination. To the quantity of urine named an ounce of a saturated solution of acetate of lead is added. Filtration is performed, and to the filtrate ammonia is added till it no longer produces a precipitate. The amount of acetate of lead employed not only suffices for the precipitation of the lithic acid, &c., but enough remains afterwards to carry down the sugar in the second precipitation with the ammonia. This precipitate is collected and washed, and then decomposed with sulphuretted hydrogen. For the determination of the sugar, the gravimetric process with the copper liquid—which I shall have especially to speak of further on—is employed.

The following are representations of the amount of sugar given by specimens of urine which, tested ordinarily with the copper solution, afforded the reaction specified in the table. It is convenient to represent the sugar in parts per 1000 for the sake of uniformity with the results that will hereafter be given relative to the blood and liver.

Sugar contained in normal urine.

	Sugar per 1000 parts.
Urine from healthy person, giving no reaction when tested ordinarily	0·276
Ditto ditto	0·206

Sugar per 1000 parts.

Urine from healthy person, giving
no reaction when tested ordinarily   0·096

Urine from patient with tubercular
meningitis   .   .   .   0·232

Urine from phthisical patient, giving
trace of reaction, tested ordinarily   .   .   .   0·433

Urine from dyspeptic patient; very
slight reaction, tested ordinarily   0·533

Urine passed after the administration of chloroform; decided reduction of copper solution, tested
ordinarily   .   .   .   1·429

In order that the bearing of these figures may be
realised, I will furnish upon the same scale the
amount of sugar that would be contained in samples
of diabetic urine of different degrees of intensity.

Sugar in diabetic urine.

Sugar per 1000 parts.

Urine of sp. gr. 1040, with 40 grains
of sugar to the fluid ounce (a
common condition in a severe
case uncontrolled by diet)   .   87·90

Sp. gr. 1040, and 30 grains to the
fluid ounce   .   .   .   65·92

Sp. gr. 1035, and 20 grains to the
fluid ounce   .   .   .   44·16

Sp. gr. 1025, and 10 grains to the
fluid ounce   .   .   .   22·27

The fact, then, that we have to deal with is that
there is no abrupt line of demarcation or distinction
of an absolute kind between the urine of health and
that of diabetes.   There is a difference, and a very
marked one, it is true, from a quantitative point of
view, at the two extremes, but the transition is not
abrupt from the one state to the other.   Every
grade of variation in the amount of sugar is encountered upon different occasions.   A sufficiency
to give a slight reaction under ordinary testing
is not uncommon, and sometimes an amount exists
that can be definitely determined without the aid of
precipitation by the lead process.   I have, from time
to time, come across specimens of urine that have
contained a few grains to the fluid ounce—say from
five to eight parts per thousand—as an incidental
occurrence.   These instances occasionally happen
without the urine, before or afterwards, presenting
a similar state, and without anything to point to
the existence of diabetes.   Such a condition has no
clinical significance, and it is only where sugar
persistently exists, or appears in the urine under
the consumption of an ordinary amount of starchy
and saccharine food, that it can be considered that
a distinctly morbid state prevails.

Having said thus much with regard to the urine,
I propose to pass to the blood.   Further on I shall
have more to say upon the urine in relation to the
blood.   I wish here simply to fix your attention upon
these facts—viz., that sugar exists in healthy urine,
that the amount can be expressed in definite terms,
and that the difference between health and diabetes,
so far as the urine is concerned, is in reality one of

degree and not of kind. If the small amount of sugar which is beyond, or just within, the reach of ordinary testing to reveal, has no clinical significance, it has, on the other hand, I consider, great importance, as I shall hereafter show, with reference to considerations which stand at the basis of our views on the relations of sugar in the system.

The knowledge obtainable from the urine is limited to that bearing on the outgoing from the body. This is all that the examination of the urine can afford. It cannot unfold to us the nature of the internal phenomena giving rise to the escape of sugar that occurs in diabetes. For the discovery of these we must seek deeper, and learn what we can of the conditions existing within the system appertaining to sugar.

The contrast is very great in our position for prosecuting research at the present day as compared with that of former times. It is curious to note, that although what may be styled a wonderfully exact description of the symptoms and course of diabetes was given by so early an authority as Aretæus, who is supposed to have lived in the first century, yet it was not until two centuries ago that the cardinal feature of the disease—the presence of sugar in the urine—was alluded to. Our renowned former associate of the College, Dr. Thomas Willis, was the first to comment upon the sweet character of diabetic urine, and to suggest that it was due to the presence of sugar. Nothing, however, in Willis' day was known about the true nature of sugar; and, strange as it may read now, it seems to have been looked upon as consisting of a

combination of sulphur and saline matter. Doubt-
fully accepted or ignored, Willis' discovery for a
long time attracted no attention. Another century,
indeed, was allowed to elapse before any further
progress was made. Dr. Dobson, of Liverpool,
towards the latter part of the eighteenth century,
initiated the prosecution of rational inquiry into the
nature of the disease. He confirmed Willis's observation, and verified his conclusions by the evidence
of vinous fermentation, and the character of the
residue from evaporation of the urine. The first
really chemical test employed for the detection of
sugar in diabetic urine was suggested by Cruik-
shank, and consisted of the action of nitric acid
in converting it into oxalic acid, which was recognised by its crystalline form. What a contrast
between the means now at our disposal and those
when taste had to be appealed to for assistance in
diagnosis, as history tells us it for some time
had by our predecessors: and even later on, when
such a test was employed as the one just referred to.

Dobson affirmed that sugar existed in the blood of
diabetics, as well as in the urine; but he relied upon
the property of taste. Rollo also asserted that he
obtained more oxalic acid from diabetic than from
healthy blood after treatment with nitric acid. The
presence of sugar, however, in diabetic blood was
afterwards disputed till Ambrosioni and Maitland
succeeded in separating it, and M'Gregor in obtaining evidence of fermentation with yeast.

Perhaps there is no disease the elucidation of
which has stood so long impenetrable, and none in
connection with which so many theories have been

from time to time broached, as diabetes. It may be almost said to have occupied the position of a kind of privileged subject for speculation. Before the recognition of sugar in the urine as a feature of the disease it was impossible that any propounded view regarding its nature could be worth consideration. When this discovery was made, it is not surprising that the kidney should be looked to as constituting the seat of diseased action ; but when it became known that the sugar existed, preformed, in the blood, there was no longer ground for assigning to this organ anything more than the performance of an eliminative operation. Faulty digestion and assimilation were now spoken of. Rollo, from observing the manner in which restriction to an animal diet checks the elimination of sugar, referred the disease to a faulty digestion of vegetable food. M'Gregor, from discovering sugar in the vomited matter from the stomach of a diabetic, after three days' restriction to beef and water, considered that the sugar was formed by a perverted action of the stomach, and from this source reached the blood.

Such was the state of knowledge upon the subject of diabetes when, about the middle of the present century, Bernard's investigations were conducted, which opened out a new field of inquiry, and from which it is not too much to say that an altogether new era has been established. Whatever may be the opinion held as to the validity of Bernard's glycogenic theory taken in the shape he has propounded it, there can be no doubt that he must be credited with a great and important discovery—a discovery which has formed the foundation of our present position, and without which it is scarcely possible that any advance could have been made.

I do not propose to detain you with an account of the experimental results upon which Bernard based his glycogenic theory. The particulars are so well known as to render it quite unnecessary that I should enter upon them here. What we want is to know how sugar naturally comports itself within the system in order that we may be in a position to ascertain the nature of the deviation from healthy action that occurs in diabetes. It is indispensable that we should be quite sure we are upon sound physiological ground before we can hope to frame well-founded pathological conclusions. If our physiological views are incorrect, we can hardly be otherwise than led astray upon entering the field of pathology.

Under the glycogenic theory, it is contended that besides the sugar derived from the starch and sugar ingested, and thus entering the system from without, the liver is endowed with a glycogenic function, through the exercise of which sugar is constantly being formed within and supplied to the circulation. It is said to be a necessary component of the physiological state that the blood should be constantly receiving this supply of sugar; and, through the glycogenic function, an animal is placed in a position of independence as regards the supply of sugar from without.

Originally the view was entertained that the sugar thus reaching the general circulation was destroyed by oxidation in the lungs. If this view were correct there would be something rational about the comportment of sugar in the system. Nothing reads more plausible than that the sugar derived from the food, and that presumed to be formed by the exercise of a glycogenic function in the liver, entering the general circulation together through the hepatic veins, should be conducted to the lungs (the organs through which the blood next passes), and become there destroyed by the influence of the oxygen of the respired air. There would be entrance at one point and destruction a step further on, and the arterial system from which the kidney is supplied would escape being charged in such a manner as to lead to an outflow with the urine; for, as I shall subsequently show, in proportion to the amount of sugar present in arterial blood so is there an elimination of sugar with the urine.

It is admitted, however, that the results of observation are not consistent with this view, and it has been long abandoned. It is now asserted that the destruction occurs in the peripheral capillaries, and chiefly in those of the muscles, but no precise or definite account is given about the manner in which the process is accomplished. We have only the surmise to deal with that it undergoes combustion resulting in the production of carbonic acid and water, and to this combustion it has been suggested part of the heat engendered by muscular contraction is attributable.

Such is the well-known doctrine comprehended under the glycogenic theory. Summarily it may be said that the issue to be dealt with is an ingress of sugar on the one hand into the general circulation through the hepatic veins from alimentation and

2

hepatic formation, and its destruction on the other hand in the peripheral capillaries. Presuming these two operations to be carried out, and to stand in such relation to each other that the destruction is equal to the ingress, there will be no accumulation within the system. Presuming, however, the relation to be altered, and the influx to be increased so as to exceed the power of destruction, or, conversely, the power of destruction to be reduced below the rate of influx, it follows as a necessary consequence that accumulation would occur. The primary phenomenon in diabetes is accumulation of sugar in the circulatory system, and the question has to be put, Which of the two factors is instrumental in determining the result? Bernard, in speaking of artificial diabetes, answers the question by saying that he is of opinion the accumulation is due to an exaggeration of the formation of sugar, and not to a lessening of its destruction.

Let us now see how the subject stands viewed by the light of the experimental evidence that can be at the present time brought forward.

It is known that several years back I obtained results of an unlooked-for nature, which gave a new complexion to the facts upon which the glycogenic theory had been based. A pupil of Bernard, and a full believer in the truth of what he had propounded, I was conducting investigations directed towards attempting to discover something about the manner in which sugar underwent destruction in the lungs, for at that time it was in these organs that the destruction was presumed to occur. In these investigations I found that what had been taken, from ordinary *post-mortem* collection of blood, to represent the physiological state, was fallacious. It had been asserted that the blood of the hepatic vein contained sugar to the extend of 10 to 15 parts per 1000 during fasting, and from 15 to 20 parts per 1000 at a period of full digestion. The mean quantity of sugar found in the liver was said to vary from 15 to 20 parts per 1000. In opposition to this I found that blood taken from the right side of the heart by catheterism during life contained what I described as a trace of sugar. It gave the same kind of reaction as the blood previously collected from the carotid artery. I could discover no appreciable difference of behaviour between the two. On subsequently obtaining blood, however, from the right side of the heart of the same animals in an ordinary manner after death it showed a presence of from 5 to 9·40 parts of sugar per 1000, and such was the state that had previously been taken as representative of the condition existing during life. The livers of the animals employed, examined a short time after death, were found to contain sugar to the extent of from 24 to 41 parts per 1000.

In three subsequent quantitative analyses of blood collected from the right ventricle during life the figures yielded, corresponding with what I designated a trace of sugar, were 0·47, 0·58, and 0·73 parts per 1000. It will be seen later on that these figures, which I obtained twenty years back, closely accord with the results I have recently procured by a new and delicate process that I shall subsequently have specially to refer to.

After noticing the fallacy of drawing a conclusion of the natural or *ante-mortem* state of the blood from

the examination as hitherto conducted, I looked to
the liver and employed means to ascertain if a
similar kind of error had prevailed with respect to
this organ. I need not refer to the details of the
processes adopted. It will be sufficient here to say
that on placing the liver instantly after death in a position to prevent a *post-mortem* production of sugar,
a totally different state from that which (through
the results upon which the glycogenic theory had
been framed) had been presumed to exist, was met
with. Instead of, as had before been believed, the
organ containing in its *ante-mortem* or natural condition a notable amount of sugar, evidence was furnished that at the outside only traces of the principle
were present. I thought at the time that the method
of analysis at our disposal was not adequate to the
indication in a reliable manner of the precise amount,
and I therefore simply spoke of it as a trace. We
shall see at a later period, from my new process of
analysis, what this expression actually corresponds
with.

My results were first communicated to the Royal
Society in 1858, and afterwards published in the
'Philosophical Transactions' in 1860. I had also
an opportunity of showing them in this lecture-
room on delivering the Gulstonian Lectures in
1862. They have been confirmed by Schiff and
Herzen, Meissner, Jaeger, Pflüger, Ritter, and Dr.
Robert McDonnell of Dublin. Flint and Lusk, in
America, admit my representation as regards the
condition of the liver to be correct, but are of opinion
that a glycogenic function exists, and that the sugar
is carried away into the circulation as fast as it is

formed. Bernard, until the last few years, remained silent upon the point, and now has given a new series of results bearing upon both the blood and the liver, obtained by an altered method of analysis. To these results it is necessary that I should direct your special attention. Everything depends on our having correct data to proceed upon, and I think I shall be able to show and convince you—in the first place, that the process of analysis that has been adopted is fundamentally fallacious; and in the next, that the condition actually existing is very different from what has been represented.

I take his description from the *Comptes Rendus* of the Academy of Sciences of Paris. The process involves the employment of potash in the quantitative determination to an extent to prevent the precipitation of the suboxide of copper. When a pure solution of sugar is dealt with no amount of potash will interfere with the precipitation of suboxide from the copper solution. In the presence of ammonia, however, the suboxide is held in solution, and the effect is simply a decoloration of the liquid. Now, in Bernard's process the organic matter existing in the product to be examined supplies the source of ammonia, and thus, instead of the red oxide falling, as in the usual method of procedure, and leading to decoloration from precipitation of the copper, the reduction occurs in the liquid, and, in proportion as it takes place the blue colour of the original solution fades, to be ultimately replaced by a colourless condition. This process is recommended from the facility it offers for enabling the precise point when decoloration is effected to be ascertained—a condition

which is often by no means satisfactorily to be determined when the ordinary process is employed, owing
to the precipitated suboxide being diffused through
the liquid and obscuring the loss of blue colour.

In the application of the process to the determination of the amount of sugar in blood the usual
method of getting rid of the albuminous and colouring matters by the aid of heat, after the addition of
sulphate of soda, is adopted. There is a minor point
which I criticised in referring to the process in a
communication presented to the Royal Society in
June, 1877. It bears on the instructions given for
the calculation of the volume of liquid obtainable
from the weight of blood and sulphate of soda employed. Amongst my remarks I spoke of the loss
of liquid by evaporation during the process of heating as calculated to produce a variation in the result.
I am glad to avail myself of this opportunity of saying that, owing to an omission in the written extracts
of Bernard's communication in the *Comptes Rendus*
it escaped me to notice that directions were given
for counterbalancing this loss by the supply of water.
So far, therefore, my criticism was unjustified, but
the main point upon which the process is open to
objection, and which I shall mention later on, remains
unaffected.

The sugar is estimated in the prepared liquid by
dropping it into 1 c.c. (about 17 minims) of the
standard copper solution, to which 20 to 25 c.c. of a
concentrated solution of potash have been added,
and noticing the quantity required to produce decoloration. The amount of standard solution taken
is very small, and any error occurring becomes

correspondingly multiplied in calculating the proportion of sugar per 1000 parts. We need not,
however, quarrel with this part of the procedure,
but assume that the exact amount of standard solution is taken, and the exact amount of liquid required
to decolorise is ascertained; 1 c.c. of the standard
solution being equivalent to five milligrammes of
sugar, this amount of sugar will be put down as
present in the quantity of liquid used to decolorise.

Through the aid of this process Bernard estab-
lishes the proposition that sugar is never wanting in
the blood of man and of the lower animals during
life, whether in a normal or in a pathological state,
and that in the normal state the amount fluctuates
between the limits of 1 and 3 parts per 1000. No
matter what the nature of the alimentation, whether
the animal be herbivorous or carnivorous; no matter
whether the blood be taken at a period of digestion
or fasting, or even during the existence of fever, it
always contains about the same quantity of sugar.
Below 1 per 1000, nutritive action, he says, is not
being carried on to its full functional extent; while
above 3 per 1000 the limit of the capacity of the
blood is passed, and the sugar overflows through the
renal organs, and the animal becomes diabetic. His
results further represent a marked difference in the
amount of sugar contained in arterial and venous
blood, the difference harmonising with the occurrence of a process of destruction in the systemic
capillaries. Taking the several analyses mentioned
(*Comptes Rendus*, t. lxxxiii, No. 6, p. 373), the mean
difference between arterial and venous blood stands
at about 0·300 part per 1000. In the experiment

furnishing the least difference, the figures stand as 1·100 per 1000 for the arterial, and 1·080 for the venous blood. In the experiment showing the greatest difference, the figures are 1·510 per 1000 for arterial, and 0·950 for venous blood, which, if true, would represent a loss of sugar amounting to 0·560 per 1000 during the passage of the blood from the carotid artery to the jugular vein, for in this instance these were the vessels from which the specimens of blood were obtained.

Now, looking at these results as they stand, it must be admitted that they give decided support to the glycogenic theory. The question, however, has to be put, Are these results to be relied upon—do they afford a representation of the true state existing? This is the point to which I will next proceed to direct your attention.

The process of analysis which has been adopted by Bernard differs, as I have mentioned, in principle of action from that customarily put into operation, the reduced oxide of copper being kept in solution by the influence of a great excess of potash on the organic matter present, instead of being allowed to be thrown down. I have tried the process against the precipitation of the suboxide and the estimation by weight of the copper in a manner I shall afterwards describe. The two processes give very discordant results, as will be seen from the table to which I will direct your attention.

Comparison of Bernard's Volumetric with Pavy's Gravimetric Process of Analysis.

Blood.	Sugar per 1000 parts.		
		Pavy's gravimetric process.	Bernard's volumetric process.
		Mean.	Mean.
I. From bullock slaughtered by Jewish method	(a) ·589 (b) ·588	·588	·975
II. Ditto ditto ditto	(a) ·510 (b) ·489	·499	(a) ·609 (b) ·640 } ·624
III. Ditto ditto ditto	(a) ·515 (b) ·535	·525	1·025
IV. Ditto ditto ditto	(a) ·698 (b) ·709	·703	·869
V. From bullock slaughtered by poleaxe	(a)1·091 (b)1·097	1·094	(a)1·311 (b)1·403 } 1·357
VI. From dog	(a) ·795 (b) ·801	·803	·800
VII. From sheep	(a) ·509 (b) ·526	·517	·761
VIII. From case of severe diabetes (obtained by cupping)	(a)4·990 (b)4·951	4·970	(a)5·000 (b)4·705 } 4·852

In each case my own process was checked by taking two samples of the blood for analysis. The same was also done in three instances with Bernard's volumetric process.

On looking at the figures, it is evident that there must be something radically wrong either with Bernard's process or that dependent for action on the precipitation of the suboxide. One or other it is quite certain must be seriously faulty. Let us stop and subject Bernard's process to the test of experimental examination before I proceed further in the matter.

In a communication read at the Royal Society in

June, 1877, I referred to the danger of relying upon results where the influence of organic matter and a large excess of potash was brought into play in the analysis. I then merely pointed this out and urged it upon argumentative grounds. I was not aware at the time that the danger suggested could so easily be shown to have a real existence. Lately, however, I have learnt that—I think I may go so far as to say— a demonstration can be afforded that the potash and organic matter lead to the development of a reducing substance, which without the presence of sugar, pro- duces a decoloration of the test. I do not ask you simply to take my word upon this matter. I will proceed to render the truth of the assertion evident to you.

I have here a liquid which has been prepared from dog's blood in the ordinary way with the use of sulphate of soda. This blood to start with con- tained only a small amount of sugar, and had been kept for two days. Besides this, whatever sugar may chance to have been present has been destroyed by boiling with a little potash. It is a well-known property of grape-sugar that it is easily destroyed by boiling for some time with a caustic alkali. I may show this action by applying the copper test to the contents of these two tubes. In the one case we have a solution of grape-sugar, and in the other this same solution after having been subjected to the process of boiling with potash. The one gives a copious precipitate of reduced oxide, the other undergoes no change. I may assume that there is no sugar in this liquid which has been ob- tained from the blood, but I will test it in the ordi-

nary way with the copper liquid, and it will be seen
that a verification is afforded by an entire absence
of reaction. I will now apply Bernard's method of
using the copper test. 1 c.c. of Fehling's solution
has been placed in this flask, with 25 c.c. of a concentrated solution of potash. The contents of the
flask are brought to the boiling point, and the prepared liquid from the blood dropped in. The blue
colour soon begins to fade, and now complete decoloration has occurred. Here we have a result
upon which the quantitative estimation of sugar
has been founded without any sugar present to
produce it.

Shortly after my communication was published in
the Proceedings of the Royal Society, a criticism
appeared from the pens of MM. Vidau, Dastré, and
d'Arsonval in the *Gazette Hebdomadaire* and the
Progrès Médical of Paris. As the last authority
signs himself "Préparateur au Collége de France,"
I imagine that what he says may be taken as agree-
ing with, if not directly inspired by, Bernard's
views. I will take no notice of the show of per-
sonality which pervades a portion of the criticism,
but feel it incumbent upon me to deal with that which
immediately concerns the point before us.

I am accused of not having rightly comprehended
the nature of the process I have criticised. I must
throw this accusation back upon my critics. The
true principle of action of the process appears to
have escaped their recognition. It does not seem
to have occurred to them that the reason of the suboxide of copper remaining dissolved is attributable
to ammonia. Dr. d'Arsonval speaks of the organic

matter which escapes coagulation in the preparation
of the blood for analysis being instrumental, in the
presence of the alkali, in maintaining the suboxide
dissolved. The organic matter is necessary, but the
observations I have made I consider justify my saying that it is *not as organic matter that it acts upon
the suboxide, but as a source of ammonia, which is the
real solvent principle.* Chemists are well acquainted
with this action of ammonia, but I do not find anything said in chemical works about a solvent influence
being exerted over the suboxide of copper by organic
matter in the presence of an alkali, as we all know
to be produced, at least by the influence of certain
forms of organic matter, in the case of the oxide.

M. Vidau records an experiment which he adduces
as showing that a fresh concentrated solution of
potash acts as a solvent of the suboxide of copper.
He says if a dilute solution of honey be dropped
into Bernard's test (1 c.c. of Fehling's solution, 10
grammes of caustic potash, and 20 c.c. of distilled
water), instead of the ordinary reaction attended
with red precipitation being perceived, the suboxide
remains dissolved, and a colourless liquid is produced.
M. Vidau here takes a form of sugar which is mixed
with extraneous organic matter, and it is through the
influence of this extraneous organic matter that the
result he describes is obtained. If he will proceed
further with his observations he will find that with
pure sugar no such results occur. If, for instance,
some of the ordinary purified cane (loaf) sugar be
tranformed into inverted sugar by boiling with a
mineral acid, and this be used, an instantaneous precipitation of the suboxide is seen. In the case of

commercial grape sugar there is sufficient nitrogenous impurity present to delay precipitation for a
moment, but it then occurs; whilst in the case of
honey a still further delay is noticed: but only this,
for precipitation here also soon occurs—that is, when
the ammonia, which the nitrogenous impurity can
generate through the influence of the concentrated
boiling solution of potash, has been dissipated.

Thus the small quantity of extraneous matter
existing in honey has been a source of deception to
M. Vidau, and led him to commit himself to an
assertion which proves to be untrue.

At the risk of wearying you I am obliged to enter
into these chemical details. It is of vital importance
that we should know the real value to be given to a
process which has been advanced by so great an
authority as Bernard for supplying information
which stands at the basis of the question before us.
I have spoken of the principle of action of Bernard's
new application of the copper test; I have suggested
that it is not through organic matter as such, but
through the influence of ammonia, developed by the
effect of the concentrated solution of potash upon it,
that the special result noticeable is produced. Any
source of ammonia will answer the same purpose,
and give rise to identically the same issue. A salt
of ammonia—for instance, the chloride of ammonium,—or a simple organic product like urea, which
is susceptible of conversion purely into carbonate of
ammonia, added to a saccharine solution, will lead
to decoloration with non-precipitation, in contact
with Bernard's test. Let me now proceed to show
what is to be said in answer to Dr. d'Arsonval upon

the question of the validity of the test in relation to
the quantitative determination of sugar.

Dr. d'Arsonval boldly and plainly advances the
proposition that "for the method to be free from
error it is necessary—first, that the decoloration
should be due to sugar; secondly, that it should be
due to sugar only." He continues, "A very simple
experiment shows that this is the case. First, leave
some blood in a vessel for thirty-six or forty-eight
hours; at the end of this time all the sugar is destroyed. The liquid derived from its treatment
with the sulphate of soda does not decolorise the
smallest trace ('le plus petit atome') of blue liquid.
Secondly, add to another portion of the same blood
a minute quantity of glucose; treat it like the first;
it decolorises the blue liquid. Add known quantities
of glucose; you will always find them to the extent
of nearly a hundredth." Dr. d'Arsonval sums up:
"The sugar is truly then the only decolorising
agent." And afterwards, when speaking of the
reason of the suboxide remaining dissolved, he says
that it is owing to the presence of the small quantity
of organic matter which has escaped the coagulating
action of the treatment with sulphate of soda, and
that this organic matter is *itself* altogether devoid of
action upon the blue liquid.

It is totally incomprehensible to me how such
assertions as these could have been made. They
are diametrically opposed to the results obtained in
my own laboratory in connection with the blood,
and to the collateral evidence that will be adduced
when I come later on to speak of the liver. I have
already given an ocular illustration relative to dog's

blood that had been kept for forty-eight hours. The product, even after being boiled for some time with a solution of potash for the purpose of adding force to the experiment, neatly and completely decolorised Bernard's test.

Here are other examples from recent observation affording similar and still more striking testimony.

A specimen of bullock's blood after slaughtering by the poleaxe was analysed at once, and found to indicate according to Bernard's process, the presence of 1.059 per 1000, and, according to my own gravimetric process, 0·840 per 1000 of sugar. It was allowed to stand in the laboratory, and on the fourth day was subjected to examination. Bernard's process of preparation, as described by Dr. d'Arsonval, was strictly followed—that is to say, equal weights of blood and of sulphate of soda were taken, and after coagulation by heat the loss by evaporation was restored by the addition of distilled water. 1 c.c. of Fehling's solution with 20 c.c. of a concentrated solution of potash was decolorised by 8·1 c.c. of the product, which, according to the correct formula for calculation taken from the discordant expressions of it given by Bernard, Dastré, and d'Arsonval, is equivalent to an indication of ·987 per 1000 of sugar. On the fifth day 9·9 c.c. were required to decolorise, representing ·808 per 1000 of sugar. On the ninth day the blood smelt strongly of putrefaction. Decoloration was in great part, though not completely, effected by 36·2 c.c. of the product, which was all that had been obtained, as filtration only without compression of the coagulum had been performed.

A corresponding observation was made on a specimen of sheep's blood obtained on the same day and afterwards allowed to stand by the side of the other.

On the first day Bernard's process indicated 0·904 and my own 0·454 per 1000 of sugar. On the fourth day Bernard's test was decolorised by 25·2 c.c. of the liquid yielded by the blood, which affords an indication of 0·317 per 1000 of sugar. On the fifth day the quantity required to decolorise was 26·4 c.c., representative of 0·303 per 1000 of sugar; and on the ninth day, when the blood smelt strongly of putrefaction, 28·2 c.c. completely decolorised. Although in this state of putrefaction, if the indication of the test were relied upon, it would have to be said that the blood contained 0·283 per 1000 of sugar.

In another similarly conducted experiment a specimen of bullock's blood, examined immediately, indicated by my own process the presence of 0·804 per 1000 of sugar. By Bernard's process the 1 c.c. of Fehling's solution was decolorised by 5 c.c. of the product, which represents 1·600 per 1000 of sugar. On the following day the blood, tested ordinarily, gave no reaction of sugar, but by Bernard's test 6·1 c.c. $= 1·311$ per 1000 of sugar, decolorised. On the third day 9·8 c.c. $= 0·810$ per 1000 of sugar, and on the fifth day, when there was a decided smell of decomposition, 10·8 c.c. $= 0·740$ per 1000 of sugar, decolorised.

The companion specimen of sheep's blood obtained at the same time contained, according to my own process 0·538 per 1000 of sugar; by Bernard's process 8·9 c.c., equivalent to 0·898 per 1000 of sugar decolorised. On the second day the blood was tested ordinarily, and gave no reaction, but 9·4 c.c. $=$ 0·851 per 1000 of sugar, decolorised. On the third day 21·3 c.c. $=$ 0·375 per 1000 of sugar, and on the fifth day, when there was decided evidence of incipient decomposition, 25 c.c. $=$ 0·320 per 1000 of sugar, decolorised.

Looking at these results, it will not be wondered at that I should speak of d'Arsonval's assertion as perfectly incomprehensible to me. In each case, by decoloration is meant that the contents of the flask were brought to as colourless a state as clear water; and to render it incontestable that the result was due to reduction, a few drops of peroxide of hydrogen were upon several occasions added, with the effect of immediately restoring the blue colour by re-oxidation.

In these last experiments an exact comparison is afforded between the amount of sugar indicated by Bernard's and my own gravimetric process. In the first set of experiments showing the comparison, which I represented in a tabular form, the figures derived from the application of Bernard's process are a little higher than they should be on account of the omission to replace, from the cause I have already explained, the loss by evaporation during the preparation of the blood for analysis by distilled water.

From what I have adduced, the inference to be drawn appears to me to be that the concentrated solution of potash, forming a part of Bernard's test, leads to the splitting up of some kind of nitrogenous

3

principle or principles present into ammonia and a reducing substance. These two factors—ammonia and the reducing substance—it may, I consider, be said, are necessary for the decolorising effect, and it would seem that the two may be developed under the influence of the potash and give a certain amount of the reaction which has been by Bernard attributed as due to sugar.

Having thus spoken of Bernard's new method of procedure in relation to the blood, I will now examine what he says in relation to the liver.

There is a cardinal point that we must keep before us as regards this organ—namely, that it contains a principle (the amyloid substance, or glycogen as it is known through Bernard) which is exceedingly prone, as one of its chemical properties, to pass into sugar. This transformation takes place with such rapidity after death as to have formerly given rise to the spread of an erroneous conclusion regarding the state of the organ during life in relation to sugar.

Many years ago I pointed out that what had been taken, from an ordinarily conducted examination after death, as representative of the condition belonging to life, was untrue. I found that if the liver were placed at the moment of death in a condition to prevent a *post-mortem* formation of sugar, a very different state was met with from what had been previously supposed to exist. For preventing the *post-mortem* change, I recommended that the piece of liver to be examined should be plunged, as speedily as possible after death, either into a quantity of boiling water or into a freezing mixture of ice and salt. The effect of boiling water is to rapidly coagulate and destroy the ferment which is the source of the transformation; and that of the freezing mixture, to suspend the occurrence of change by virtue of the reduction of temperature. Thus treated, the liver was found to contain only a trace of sugar, instead of a notable amount, as had been previously believed. Before these experiments were performed no idea had been entertained that the result of an ordinarily conducted examination after death furnished a fallacious representation, and what I described was soon confirmed by various authorities in this country and abroad.

With reference to these results, Bernard remarks, in the '*Comptes Rendus*' for May 28th, 1877:—"All the results that have been obtained, which appear contradictory to the existence of a notable amount of sugar in the liver during life, are defective and imbued with error. These remarks apply to the experiments of Pavy, Meissner, Ritter, Schiff, Lusana, &c. These experimentalists have not proceeded with sufficient precision. They have not given the relative quantities of water and liver employed, have not determined the proportion of sugar found, and have often made use of processes too rough to permit an amount of from 1 to 2 per 1000 being recognised in the liver."

To rectify the alleged error, and to represent the amount of sugar contained in the liver during life, Bernard has recourse to the following extraordinary process :—A porcelain capsule is placed on a support resting on the scale of a balance, and arranged so that a jet of gas can be made to play underneath. Sixty grammes (about two fluid ounces) of water are placed in the capsule, and when brought to the boiling-point are counterpoised by weights in the opposite scale. At this stage the piece of liver to be examined, weighing about twenty grammes, is plunged into the water, and the weight taken, with the jet of gas playing under the capsule during the while. The heat is continued, and boiling maintained for five to ten minutes. The liver can now be taken out, pounded in a mortar, and transferred, with the water in which it was boiled to a beaker. The contents of the beaker being again boiled, the liquid is ready to be separated by straining and squeezing through muslin, and has to be brought to a known volume. The proportion of sugar is then determined by Fehling's solution.

From the application of this process, Bernard asserts that the liver during life contains sugar to the extent of from 1 to 3 parts per 1000.

I will not say anything about the awkwardness of the part of the process in which the weighing of the liver taken is concerned. The process itself involves a total misconception of the precautions necessary to obtain a representation of the condition existing at the moment of removal from the animal. It is known that the liver contains both amyloid substance and ferment, and that the former is transformable into sugar by the action of the latter with a rapidity proportionate to the height of the temperature short of the ordinary coagulating point of albuminous matter, which is destructive of the active property of bodies of the nature of ferments.

If it be required, through the influence of heat, to prevent *post-mortem* change, the piece of liver should be plunged into a considerable quantity of boiling water—at least about a quart—in order that the heat may penetrate it and destroy the ferment as rapidly as possible. With a piece of liver twenty grammes in weight immersed in only sixty grammes of water the relation is such that the temperature of the water is for a short time brought down, and the coagulating process correspondingly retarded. Under these circumstances, the process of transformation is actually favoured in the part to which coagulation is approaching. That what I have stated has no imaginary foundation I am able to show by appeal to observation.

Here are the results obtained after submitting the same livers instantly after death to Bernard's plan of preparation for analysis and my own plan by freezing, which I shall describe hereafter. The same process of analysis—viz. the gravimetric process, also to be hereafter described—was employed for each, so that it was only the mode of preparation for the analysis that differed.

Amount of Sugar found after Bernard's and the Freezing Methods of Preparation for Analysis.

	Sugar per 1000 parts.	
	Bernard's method.	Freezing.
Liver of cat .	. 1·216 .	. 0·345
Do. do. .	. 0·850 .	. 0·056
Do. rabbit	. 0·869 .	. 0·069

I have stated that these results were obtained, as regards the actual estimation of the sugar, by my own process of analysis. If Bernard's method of procedure is adopted throughout, much higher figures are given. He simply states in the ' *Comptes Rendus*' that in obtaining his figures the estimation was made with Fehling's solution. From what he says, however, about the position held by his new Fehling's solution and potash process, I take it this was adopted for the liver as well as the blood, and, if so, the remarks I have already made will apply here, and even, I learn from observation, with greater force than in the case of the blood.

There are grounds for asserting that the process gives in a marked manner the reaction of sugar when no sugar is present. It is admitted that the liver of the human subject, after death from disease, is usually found free from sugar. Now, I have procured specimens of liver from the *post-mortem* room, and noticed that, although they have given no trace of reaction on being tested ordinarily with the copper solution, yet by Bernard's process decoloration from reduction of the oxide of copper has been readily effected.

Here are liquids obtained by treating the liver with sulphate of soda and heating to procure a clear product. The livers behaved in the same way to begin with, but they were purposely kept four and three days respectively before the products were obtained, in order to render the evidence more conclusive. One was yielded by a case of death from suppurative peritonitis, and the other from malignant jaundice or acute yellow atrophy of the liver. Although, as may be seen, giving not the slightest sign of reaction on being tested ordinarily, yet it only requires under 15 c.c. to completely decolorise 1 c.c. of Fehling's solution treated with

20 c.c. of the concentrated solution of potash, according to Bernard's plan.

Here is another case of a liquid similarly obtained and boiled for some time with a little potash to secure the destruction of sugar, presuming any to have been originally present. You observe, on Bernard's process being applied, that the test in the flask becomes decolorised after a certain amount of liquid has been dropped in.

Not only is this behaviour noticeable with the liver presumably free from sugar, but with liquids obtained from other organs. I have examined the spleen, kidney, and muscle in relation to this point, and found that each affords a behaviour indicative, according to Bernard's test, of the existence of a notable amount of sugar. Whether procured from the *post-mortem* room, the physiological laboratory, or slaughterhouse, if these organs are treated with sulphate of soda, as mentioned for the liver, the liquid product furnished exerts a decolorising action upon the potash and Fehling's solution test. Hence if this reaction were relied upon, it would have to be said that these organs contained a notable amount of sugar, although, as is well known, it is asserted as a part of the glycogenic doctrine that the liver alone of the organs of the body contains sugar.

Healthy urine also produces a strong decolorising effect, and not only urine in the fresh state—which I have shown contains a certain minute amount of sugar—but urine which has been kept for a considerable time. A short time since I purposely kept some urine, which, tested ordinarily, gave no

reaction to begin with, for nearly three weeks; and
although, at the end of that time, there was a
mouldy growth upon the surface, a small quantity
sufficed to remove the colour from Bernard's test.
I have even kept a specimen of urine derived from
a patient suffering from hæmaturia until it had
become quite putrid; and this also, after coagulation and filtration, produced the same effect.
It may fairly be assumed, under such circumstances, that no sugar could exist. The reducing
action of the uric acid present in the urine
appears to me inadequate to account for the result.
Moreover, in some observations which are not yet
sufficiently complete to permit me to do more than
cursorily allude to here, I have, with the employment of chloride of ammonium to furnish the
required ammonia, obtained results suggestive that
the extent of reducing action is influenced by the
amount of potash used.

From the array of evidence which has been produced, the only conclusion that appears to my
mind permissible is that Bernard's new test is
fundamentally fallacious. If the test is faulty, the
results furnished by it require to be dealt with
accordingly. The authority given by the weight
of Bernard's great name renders it necessary that I
should speak plainly on this matter, and it is my firm
conviction that the ground must be cleared of the
deceptive evidence that has recently been adduced
to rightly fit us for approaching the investigation of
the pathology of diabetes.

Impressed with the importance of being able to
express in precise and reliable terms the amount

of sugar corresponding with what in my original experiments I described as a trace in relation to the blood and the liver, I have given my attention to the application of the copper test gravimetrically instead of volumetrically. It is now a common practice with chemists in sugar determinations to employ the former method of procedure in preference to the latter. Greater precision and certainty belong to a result obtained by the balance than to one derived from watching the gradual disappearance of colour, and mentally deciding when the required point is just, and only just, obtained, as in the ordinary volumetric process. Moreover, in the case of minute quantities of sugar, as when the blood and the liver in a state belonging to life are dealt with, the suboxide is thrown down in so finely divided a form as to remain diffused through the liquid, and obscure the determination. Indeed, under these circumstances, the difficulty is such as to deprive the estimation of the desired authority as an exact expression of quantity. Hence the reason that actual quantitative results have not hitherto figured more frequently in the discussion of the physiological relations of sugar.

In adopting a gravimetric process, the copper may be weighed in different states. The suboxide may be collected and weighed as such, or converted into the oxide and the weighing of this undertaken. From the difficulty, however, that exists in obtaining the metallic oxide in a pure and uniform state, some want of precision belongs to the results thus derived. Looking for something more susceptible of delicate exactness, the plan suggested itself of dissolving the precipitated suboxide and throwing the copper down in the metallic form by galvanic action upon a platinum surface, in the same manner as is now extensively done in the assaying of copper ores. Several difficulties at first presented themselves in the execution of this process, and at one time I almost thought it would have to be given up; but one by one they have been overcome, and now I feel that, with the requisite care in manipulation, it supplies a means of furnishing reliable information. The close agreement noticeable in the results yielded by counterpart analyses gives strong testimony in favour of the process being susceptible of great precision.

The object to be attained is to determine the amount of copper which corresponds with the reducing effect of the sugar existing in the product analysed. We must first prepare a suitable liquid for the application of the copper test solution. The precipitated suboxide has then to be collected in a separate form; it is next dissolved, and the copper thrown down by galvanic action upon a cylinder of platinum foil. The final part of the operation consists in ascertaining the weight of the copper thus thrown down. Such are the principal features of the process, but I consider it requisite to furnish an account of the manner in which the several steps of it are carried out. I am anxious not to enter into unnecessary detail, but I feel that these considerations, occupying the fundamental position they do in relation to the physiological conclusions upon which our views concerning the nature of diabetes are based, should not be cursorily passed over. Unless you are satisfied that the chemical method of procedure is entitled to credit, I could not expect you to accept the conclusions derived therefrom.

Let us suppose, then, that we have a specimen of blood to submit to examination for the quantitative determination of sugar, the following is the manner in which, according to the plan proposed, the operation is conducted.

Forty grammes of sulphate of soda in small crystals are weighed out in a beaker of about 200 c.c. capacity. About 20 c.c. of the blood intended for analysis are then poured upon the crystals, and the beaker and its contents again carefully weighed. In this way the precise weight of the blood taken is ascertained. The blood and crystals are well stirred together with a glass rod, and about 30 c.c. of a hot concentrated solution of sulphate of soda added. The beaker is placed over a flame guarded with wire gauze, and the contents heated until a thoroughly formed coagulum is seen to be suspended in a clear colourless liquid, to attain which actual boiling for a short time is required. The liquid has now to be separated from the coagulum, and the latter washed to remove all the sugar. This is done by first pouring off the liquid through a piece of muslin resting in a funnel into another beaker of rather larger capacity. Some of the hot concentrated solution of sulphate of soda is then poured on the coagulum, well stirred up with it, and the whole thrown on the piece of muslin. By squeezing, the liquid is expressed, and to secure that no sugar is left behind, the coagulum is returned to the beaker, and the process of washing and squeezing repeated.

The liquid thus obtained may be fairly regarded as containing all the sugar that existed in the blood. From the coarse kind of filtration and squeezing employed, it is slightly turbid, and requires to be thoroughly boiled to prepare it for filtration through ordinary filter-paper. A perfectly clear liquid runs through, and to complete this part of the operation the beaker that has been used and the filter-paper are washed with some of the concentrated solution of sulphate of soda before referred to.

The next step is boiling with the copper test solution. The liquid is again placed over a flame, and brought to a state of ebullition. A sufficient quantity of the copper solution to leave some in excess is now poured in, and, from the time of recommencement of boiling, brisk ebullition is allowed to continue for a period of one minute. This suffices for all the sugar to be oxidised, and, accordingly, for all the action upon the copper solution to be completed. There is no risk during this time of spontaneous change occurring in the copper solution; but observation has shown that, should the boiling be continued for a lengthened period (by which I mean ten minutes or a quarter of an hour), the copper solution undergoes alteration, and no longer possesses the power of resisting spontaneous reduction. As regards the amount of copper solution to be used, although 10 c.c. of the test as ordinarily made are found to suffice for 20 c.c. of the blood of animals in a natural state, yet it is well to employ from 20 to 30 c.c. to secure that it is thoroughly in excess. Where, from any circumstance, larger quantities of sugar exist in the blood, more in proportion of the test must, of course, be used.

The precipitated suboxide of copper has now to be separated from the excess of copper solution. Experience shows that filtration through filter-paper cannot be resorted to for the purpose. In the first place, the pores of the paper tend to become blocked up and filtration to be stopped; and in the next—and this is a fatal objection—the paper absorbs and so tenaciously holds some of the copper solution that it cannot be effectually washed out. A plug of asbestos in a filter-funnel may be used instead; but asbestos varies considerably, and it is not always easy to procure it in a state to answer well for free, and at the same time perfect, filtration. A material, however, which has somewhat recently been introduced—viz. glass-wool—fully furnishes what is wanted. Properly packed in the neck of a funnel, it permits filtration to be effectively and easily performed. The filtrate should always be carefully examined, to see if the plug has been sufficiently tightly packed to keep the whole of the precipitate back. Should the crystallisation of the sulphate of soda in this or the preceding filtration interfere with the continuance of the operation, the funnel may be placed over a beaker holding some liquid kept in a state of ebullition, or heat may be applied in any other way.

The suboxide having been collected, and washing with distilled water performed, it is returned to the beaker in which the reduction was effected, to secure that none of the precipitate that may have been adhering to the sides of the vessel is lost. The plug is pushed with a glass rod from the neck of the funnel held in an inverted position over the beaker, and the funnel washed and its surface cleaned from all adhering precipitate. We have now the suboxide in a fit state to dissolve, and, until I resorted to the use of peroxide of hydrogen to effect its oxidation, a difficulty presented itself in this part of the operation, the precipitate requiring an amount of acid to dissolve it which interfered with the subsequent deposition of copper by galvanic action. After the addition of a few drops of peroxide of hydrogen, a very small quantity of nitric acid (a few drops only) is sufficient to lead to instantaneous solution, and after boiling to decompose the excess of peroxide of hydrogen, the contents of the beaker, consisting of filter-plug and dissolved precipitate, are poured into a funnel containing a loose plug of asbestos or glass-wool, to obtain the liquid in a separate form. The requisite washing with distilled water having been performed, there only remains the final stage of the process to be accomplished.

The liquid to be now dealt with contains the copper in the form of nitrate, which experiment has shown to be the most suitable for yielding a pure metallic deposit by galvanic action. For the purpose of collecting the deposit, a cylinder of platinum foil, soldered to a platinum wire for hooking on to the negative pole of the battery, is employed. This is immersed in the liquid so as nearly to touch the bottom of the vessel, and inserted within it is a spiral coil of platinum wire made to constitute the positive pole of the battery. In order to secure a good continuous connection, the platinum spiral is closely bound to the copper conducting-wire of the battery, and the other pole is provided with a platinum hook for the suspension of the cylinder. This precaution has been found necessary from the ready manner in which copper exposed ends become oxidised, and rendered imperfect conductors by the oxygen escaping from the liquid underneath. It may also be mentioned that the platinum spiral, after several days' use, acquires a brown surface, and requires to be occasionally cleaned by immersion in hydro-chloric acid. At the end of twenty-four hours' exposure to galvanic action the weight of the cylinder with the deposited copper is taken. The cylinder is lifted quickly out of the liquid, and instantly plunged, first into distilled water, and then into spirit, the latter being used to avoid the occurrence of oxidation of the copper in drying. After drying by suspension in a water-oven the process of weighing is performed, and it is hardly necessary to say that a delicate chemical balance is required for the purpose. The weight of the cylinder being known and subtracted gives the weight of the copper that has been thrown down. In the case of an analysis of blood containing an ordinary amount of sugar, and therefore yielding a limited amount of copper to be deposited, twenty-four hours have usually been found to suffice for the completion of the operation; but it is necessary there should be no uncertainty upon this point, and to secure this, the following course of procedure should be adopted. After the weighing has been effected, the deposited copper is dissolved off by immersion of the cylinder in nitric acid, and the cylinder then returned into the liquid to see if any fresh deposit occurs. If, after some hours, no copper tint is seen, the operation may be regarded as completed; but if more deposit has occurred, the immersion must be continued, and another weighing performed, and this repeated till the platinum surface remains untinted. We have here the most delicate test there is for the presence of copper, and if no appearance of deposit is perceptible it may be safely concluded that no appreciable amount of copper remains in the liquid.

The galvanic action requires to be steadily and continuously maintained, and a modification of Fuller's mercury-bichromate battery has been found to answer best for use. The arrangement that has been employed in my experiments consists of an outer cell provided with two carbon plates, and charged with bichromate of potash dissolved to saturation in dilute sulphuric acid. Into the inner porous cell a little mercury is poured, and it is then filled up with water. An amalgamated zinc rod is inserted, and dips down into the layer of mercury at the bottom. This battery, it is found, gives a steady current, and, used every day, will remain in good working order for at least a fortnight, all that is necessary being to pour out the liquid in the porous cell when it has become green from reduction of the diffused bichromate solution, and replace it with water. Attention is, of course, necessary to secure that the proper battery-power exists to effect the deposition of the copper, and when the current becomes weak, the zinc rod must be cleaned, and the bichromate of potash solution replenished.

When sugar is boiled with the copper solution the change occurring stands in the relation of one atom of the former to five atoms of cupric oxide. One atom of sugar is oxidised by, or reduces, five atoms of cupric oxide. This is the foundation of the action involved in the operation of the test, and the calculation of the amount of sugar present is made accordingly. Taking 63·4 as the atomic weight of copper, and 180 as that of glucose ($C_6H_{12}O_6$), 317 parts of copper will stand equivalent to 180 parts of glucose. Thus one part of copper corresponds to ·5678 of glucose, and in calculating the amount of sugar in the blood analysed, the weight of the copper deposited has only to be multiplied by ·5678 to give its equivalent in glucose. The quantity of sugar in the amount of blood taken for analysis being thus determined, the required information is supplied for expressing the proportion for 1000 parts.

Experience and careful consideration enable me to say that I feel that the process I have described may be confidently appealed to for giving precision to our knowledge regarding the relations of sugar in the animal system. It is true, time and delicate manipulation are required for obtaining the results, but these are matters which ought to be regarded as of no consideration in view of the object for which the process is proposed.

4

Before speaking of the results I shall have to place before you obtained by the process I have described, I will refer to the criticism of it that has been put forward by the partisans of Bernard's views in Paris. My observations were communicated to the Royal Society in June 1877, and shortly afterwards the 'Progres Médical' contained an article by M. Dastré, and the 'Gazette Hebdomadaire' one by M. Vidau, and another by Dr. d'Arsonval, *Préparateur au Collége de France.* I will not condescend to notice the hostile remarks which the latter writer has indulged in. Had he been better acquainted with the literature of the subject, he would not have committed himself to the representation she has done. Only so far as the interests of science demand will I occupy time in adverting to the criticism that has been set forth. Through mistaking the real source of the non-precipitation of the suboxide of copper in Bernard's application of the copper test, my critics have founded their chief objection to the mode of analysis I have proposed. The two processes cannot be looked upon in the same light. Because with the large amount of potash—from 6 to 10 grammes in 20 c.c. of liquid, according to the instructions given—the suboxide is held in solution, it does not follow that the slight amount of alkali present under my own mode of testing should lead to a solvent action. The fact is that in Bernard's process the suboxide does not fall, and in my own it does. I have already given my reasons for asserting that it is not to organic matter as such, but to organic matter placed under circumstances to develop ammonia, that the result, when no precipitate falls, is attributable. Unless ammonia is developed, no interference with precipitation occurs. The nitrogenous organic matter existing in the colourless liquid prepared with the assistance of sulphate of soda from blood, is evidently not in a form to readily develop ammonia; otherwise in the ordinary rough application of the copper test in a test-tube no precipitation would be noticeable, whereas observation shows that it is. In my own method of procedure a much more diluted state of the alkali even exists, on account of the presence of the liquid used in washing the coagulum and the filter. Ordinarily, for instance, from 130 to 150 c.c. of liquid are obtained, when the washing is completed, from 20 grammes of blood. Into this, say, 25 c.c. of Fehling's solution are poured. The alkali present will be diluted to the proportion of about 1 per cent. In such a state of dilution there is nothing irrational in assuming that it may fail to possess the power of severing the elements of a nitrogenous organic principle, although in the proportion of about 1 to $2\frac{1}{2}$ or 3, as in Bernard's process, it may be effectual in doing so.

Facts, however, are better than words, and, without any further argument, I will let the results of observation speak for themselves. The fact is that the organic matter present in the product for analysis from blood *does not*, to any appreciably significant extent, interfere with the deposition of suboxide, as is proved by the following results.

In the first place I tried the effect of adding known quantities of sugar to a specimen of sheep's blood which had been just collected from the animal.

An analysis was made of the blood itself to determine the amount of sugar already contained in it. Sugar was added in three different proportions—viz. 20, 40, and 80 milligrammes. The quantities found, after taking the mean of the duplicate analyses performed in each case, and after making the deduction for the sugar originally present in the blood, were 19, 40, and 79 milligrammes respectively.

An observation was also made upon the product obtained from a piece of frozen liver. Ten milligrammes of sugar were added and 9·6 milligrammes found after deducting that shown by analysis to be originally present.

It may perhaps be said with regard to these experiments that they do not afford the full proof needed, as the blood and product from the liver in themselves allowed a certain amount of suboxide to fall, and this being the case, the sugar added would not be reached by the influence of the organic matter present. To meet this objection, I have exposed sheep's blood for twenty-four hours in a situation where the thermometer stood at about 110° Fahr. (43° Cent.), with a view of effecting a complete removal of its sugar. At the end of this time the analysis gave rise to no deposition of suboxide. Two twenty-gramme portions were subjected to the ordinary preparation with sulphate of soda, and twenty milligrammes of sugar were then added to each of the products. The amount of sugar found, taking the mean of the two analyses, was 19·6 milligrammes. I consider that a complete answer is given to the allegation that the organic matter remaining in the prepared product for analysis vitiates the result by interfering with the deposition of suboxide and thus giving rise to a lower indication of sugar than should be afforded. If this action were exerted in the case of blood in its natural state, is it not reasonable to infer that the effect would also be perceptible upon some of the sugar added in the experiments I have recorded?

As to a suggested loss in the process of analysis by sugar being allowed to remain behind in the coagulum, in filtration, and in washing, I would simply tell my critics that the operations performed in my laboratory are not conducted in such a loose manner as to engender fallacy from such a cause. The proposition is a perfectly gratuitous one, and is shown to be groundless by the close accord noticeable in the counterpart analyses, to which I shall have presently to direct attention. Indeed, the closeness of some of the results is almost surprising, and might even be considered satisfactory for an inorganic instead of an organic material. I may remark further that it should be borne in mind that the figures represent parts per 1000, whilst the analysis was conducted upon only about twenty grammes of blood. The difference, therefore, in the results as they appear is fifty times greater than that obtained in the actual process of weighing.

I am not unmindful of what I said in speaking of the urine, with regard to the concealment of the reaction belonging to the minute quantity of sugar susceptible of being shown by lead precipitation to be present, and it will naturally be required of me to explain the apparent anomaly that here exists. The explanation is readily given, and what is noticeable, in fact, lends support to the view I have expressed.

It is well known that urine may be easily made to give rise to the evolution of ammonia. I need not enter into the chemical question of whether it exists ready formed or not; or, assuming its existence as there is ground for doing, in what state of combination it is present; suffice it for my purpose to say that the simple application of heat to perfectly fresh urine leads to its escape, and after the addition of an alkali heat produces a pretty free evolution distinctly appreciable to test-paper. If a little moistened test-paper be held at the mouth of a test-tube containing a mixture of urine and Fehling's solution, and heat be applied as in the ordinary mode of testing for sugar, strong evidence is afforded of the escape of ammonia. Now, ammonia is a principle which, as I have before explained, possesses the power of dissolving suboxide of copper. We have, therefore, in the urine sufficient to account for the concealment of the suboxide produced by the reducing influence of the minute quantity of sugar present. With more sugar present, the amount of suboxide formed exceeds the solvent power of the ammonia, and a precipitate falls. By looking to ammonia as the source of the obscure behaviour of the copper-test with urine containing small quantities of sugar, the different kinds of reaction noticeable upon different occasions—sometimes, for instance, more or less reduction after long boiling, sometimes no appearance of reduction till after the test-tube has been placed aside for a little while, and some-times the contents of the tube becoming yellow and translucent for a little time, and then all of a sudden throwing down a quantity of precipitate of the ordinary kind—become at once intelligible. From the position held by the copper-test as an agent so frequently made use of as it is for clinical purposes, the point I have been referring to appears to me to possess an interest beyond its connection with the question under immediate consideration.

With the product obtained by the agency of sulphate of soda from blood, a very different result occurs as regards the development of ammonia. Let a portion be placed in a test-tube and treated with Fehling's solution in the usual manner for testing for sugar, and boiled. A piece of moistened test-paper held at the mouth of the test-tube shows no decided evolution of ammonia; at least, if any at all occurs, the amount must indeed be very slight. Let a little urine be now dropped into the tube, and the contents be reboiled. A strong effect is at once produced upon the test-paper.

Thus, then, because a certain amount of sugar may escape being revealed by deposition of suboxide from the copper test in the case of the urine, it does not follow that such should occur in the analysis of blood. There is a difference in the condition of the liquids to which the test is applied; and, that there should be a difference in the result of testing is not only in harmony with what might be looked for, but helps to give consistency to the whole.

Such is the reply I have to make to the criticism of the process upon which the results I shall have to bring forward are founded. Instead of the difference between the results obtained by Bernard's and my own process being due, as suggested, to a defective action of the latter, I insist that the fault lies on the other side. I have shown that Bernard's process gives the behaviour in a notable quantitative manner belonging to sugar under circumstances inconsistent, it may be said, with its presence, and the action thus occurring will account for the higher figures obtained by it.

"I will not," says Dr. d'Arsonval, "have the cruelty, like M. Vidau, to state to Dr. Pavy that his process is found to have been described a number of years back by Fresenius and Henri Rose, whom he could have quoted." Unfounded upon reality as this assertion is, the consideration shown to me deserves some recognition on my part, and I therefore, in concluding this matter, will not have the cruelty to say more than that capacity as an observer, and fidelity as a writer, are needful qualifications for a man who professes to be devoted to science, and I leave it to be judged by others how far these qualities are to be found displayed by the "Préparateur" of the College of France in his communication to the 'Gazette Hebdomadaire' to which I have had to direct attention.

I now come to the amount of sugar shown to be present in the blood by the process I have described and discussed. I have results to place before you derived from the examination of the blood of the dog, sheep, and bullock. These results were obtained from specimens collected with the requisite precautions to afford a representation of the natural state. Observation shows that the amount of sugar undergoes a very speedy and marked alteration under deviations from a normal and tranquil condition; and to avoid exposure to fallacy, it is absolutely necessary to give as strict attention to what may be called the physiological as to the chemical part of the experiment. To escape the vitiating influences so easily arising from the effect of the operation to procure the blood during life, I consider it best that it should be collected at the moment of death. In adopting this course, the chief point to be looked to is to guard against allowing time for the contents of the circulation to become influenced by the production of sugar that occurs in the liver as a *post-mortem* effect.

Here are the results derived from my observations conducted upon dogs. In the first six, the blood was obtained by pithing the animal and *instantly* afterwards inserting a scalpel into the chest and freely incising the heart and large vessels. The chest was then quickly opened and the blood dipped out and treated for analysis before coagulation had occurred. In the seventh observation, the blood was obtained by division of the jugular vein instantly after pithing, instead of from the chest. Counterpart analyses were in each case made for the sake of verification.

Amount of sugar present in the blood of the dog.

Sugar per 1000 parts.
Mean.

Observation 1	(a) 0·743	0·751
	(b) 0·758	
,, 2 ...	(a) 0·776	0·786
	(b) 0·797	

Sugar per 1000 parts.
Mean.

Observation 3	...	(a)	0·706	0·700
		(b)	0·694	
,, 4	...	(a)	0·770	0·766
		(b)	0·762	
,, 5		(a)	0·777	0·786
		(b)	0·795	
,, 6	...	(a)	0·795	0·803
		(b)	0·811	
,, 7	...	(a)	0·932	0·921
		(b)	0·910	

Average, 0·787 per 1000.

The blood from the sheep was obtained from the slaughter-house. The animals were killed in the usual way—namely, by passing a knife through the neck and dividing the vessels, and it was the first portion which escaped that was collected for analysis. The time elapsing between the collection and the commencement of analysis did not exceed a quarter of an hour. As coagulation had taken place, the clot was snipped with a pair of scissors into fine pieces, for the sake of convenience and for securing uniformity.

Amount of sugar present in the blood of the sheep.

Sugar per 1000 parts.
Mean.

Observation 1		(a)	0·456	0·470
		(b)	0·484	
,, 2	...	(a)	0·538	0·490
		(b)	0·447	
,, 3	...	(a)	0·509	0·517
		(b)	0·526	

Sugar per 1000 parts.
Mean.

Observation 4	...	(a) 0·577 (b) 0·542	}	0·559
,, 5	...	(a) 0·590 (b) 0·548	}	0·569
,, 6	...	(a) 0·527 (b) 0·525	}	0·526

Average, 0·521 per 1000.

In the case of the bullock, I have availed myself of the Jewish method of slaughtering for obtaining the blood. Under this method the animal is killed by drawing a sharp knife across the neck and cutting through all the parts in front of the vertebral column. It is arterial blood that is thus furnished. One hour elapsed before the commencement of the analysis.

Amount of sugar present in the blood of the bullock.

Sugar per 1000 parts.
Mean.

Observation 1	...	(a) 0·698 (b) 0·709	}	0·703
,, 2	...	(a) 0·515 (b) 0·535	}	0·525
,, 3		(a) 0·500 (b) 0·484	}	0·492
,, 4		(a) 0·464 (b) 0·449	}	0·456
,, 5	...	(a) 0·510 (b) 0·489	}	0·499
,, 6	...	(a) 0·589 (b) 0·588	}	0·588

Average, 0·543 per 1000.

I have laid stress upon the importance of giving attention to the manner in which the blood is collected for examination. Killing by the poleaxe does not yield a suitable specimen, unless the subsequent incision into the lower part of the neck for the escape of blood is more expeditiously performed than is usually done. I have learnt by observation that with blood thus collected analysis may show double the quantity of sugar found under the Jewish process. A striking exemplification is here afforded of how beset with difficulty this subject is, and how guarded it is necessary to be to escape from fallacy.

The period that elapsed between the collection of the blood and its examination has been mentioned on account of what Bernard has said with regard to the disappearance of sugar in blood that is kept. I have carefully investigated this point, and am able to state that the rate of disappearance under exposure to an ordinary temperature is only slow, and that nothing of significance is perceptible within a moderate time after removal. My results were published in the ' Proceedings' of the Royal Society for June, 1877, and they show that even at the end of twenty-four hours a considerable proportion of the sugar originally present is to be found.

Having said thus much with regard to the quantity of sugar naturally existing in the blood, which we find, according to the information supplied, amounts to no more, speaking roundly, than $\frac{1}{2}$ per 1000 in the case of the sheep and bullock, and $\frac{3}{4}$ per 1000 in that of the dog, I will pass to the liver, and show what quantitative analysis reveals in relation to this organ.

In the first place, recourse must be had to a method of preparation for the application of the analytical process which will secure that a representation of the state belonging to life is given. The plan I have adopted is simple, easily performed, and cannot be otherwise than free from objection. It is this. The animal is killed by pithing, and as speedily as it can be done, the abdomen is laid open, and a piece of the liver obtained and plunged into a freezing mixture of ice and salt. The object is to lower the temperature sufficiently to check the *post-mortem* transformation of amyloid substance as quickly as possible. The piece of liver should not be very thick, and one or two incisions with the scalpel may be made into it the better to secure the rapid penetration of the freezing influence. The temperature of the freezing mixture should be below 0° Fahr., and the ice and salt should be in a partially liquefied state, in order that the most favourable condition for operation may exist. The desirability of freely stirring the piece of liver about in the freezing mixture for the first moment or two after being plunged in will naturally suggest itself.

After the completion of the freezing process, the next step is to extract whatever sugar may exist in the frozen mass without affording an opportunity for any fresh formation to occur. Whilst in the frozen state there is no danger of change, and weighing for analysis may be deliberately performed. It is convenient to take about twenty grammes, and when the weighing has been finished, the portion taken is transferred to a mortar, thoroughly pounded up, and then treated with spirit. From 300 to 400

c.c. should be used, and when once added, the subsequent steps may be leisurely carried out. Whilst
dissolving out the sugar, the spirit coagulates the ferment and leaves behind the amyloid substance. After
the pounded liver and spirit have been allowed to
digest together for some time, filtration is performed,
and the filtrate brought down to a small bulk over
the water-bath. There happens to be a resinoid
material in the spirituous extract, which interferes
with filtration after boiling with the copper test.
I have found it necessary to get rid of this, and the
addition of some sulphate-of-soda solution answers
the purpose. After the addition has been made,
evaporation is still carried on until the resinoid
matter is seen to have collected in a separate form.
It is now easily removed by filtration, but the
required washing must be performed with the
sulphate-of-soda solution.

The process has now reached the stage for the
employment of the copper solution, and the subsequent steps of the operation are carried out in
precisely the same manner as I described for the
blood. Here are the figures yielded by observations
conducted upon the cat, rabbit, and dog.

*Amount of sugar found in the liver, taken at the
moment of death.*

CAT.			Sugar per 1000 parts.
Observation 1	...	...	0·545
,, 2	...	...	0·431
,, 3	...	...	0·087
,, 4	...	...	0·182

CAT.		Sugar per 1000 parts.
Observation 5	... ...	0·248
„ 6	... ...	0·345
„ 7	... ...	0·056
RABBIT.		
Observation 1	... ...	0·554
„ 2	... ...	0·597
„ 3	... ...	0·069
DOG.		
Observation 1	... ...	0·315

This completes what I have to say about the amount of sugar naturally existing in the blood and liver during life. The figures I have placed before you will, I think, be seen fully to justify the assertion I made in my original communication upon the subject twenty years ago. Our position was then such that I was unable to give reliable quantitative determinations, and I spoke of the amount estimated from the behaviour of the copper test as corresponding with a slight, or trace of, reaction.

The next point to which I must solicit your attention is the question of destruction of sugar within the system. The ingress of sugar into the general circulatory system necessarily involves a corresponding destruction, or removal in some other way, to obviate accumulation. When the glycogenic theory was first propounded, it was thought that sugar underwent destruction during the transit of the blood through the lungs. This harmonised well with the requirements of the case. A flow of sugar from the liver, derived either from ingestion or formation, or from both; its passage along the inferior

cava and through the right cavities of the heart to
the pulmonary artery ; and its disappearance in great
part on reaching the lungs, the capillaries of which
stand next in the course of the circulation, all fitted
well together, and presented a semblance of reality.
It was through some early experiments of my own
that the difference originally supposed to exist
between the blood which had passed through the
lungs and that flowing to them was found to be
based upon fallacious evidence.   Bernard even
subsequently abandoned his view, and latterly adduced experimental results upon which he founded
the proposition that destruction of sugar to a notable
extent occurs in the systemic capillaries.   His last
view in short is that the sugar reaching the circulation, as presumed under the glycogenic doctrine,
passes through the lungs, and is conveyed in the
arterial blood throughout the system, to undergo
destruction whilst traversing the capillaries of the
muscular and other general tissues of the body.

This view is founded upon quantitative results
obtained by his volumetric process of analysis, which
I have previously criticised.   I have nothing here to
say about the results viewed from a chemical stand-
point.   Discarding the consideration of this part
of the question as having been disposed of, there
are other matters to be attended to on looking to the
results as affording a comparative representation of
arterial and venous blood under the natural conditions belonging to life.   Taking the whole experiment requiring to be conducted, there is a physiological as well as a chemical part to be dealt with,
and more error even may arise from the omission of

the proper precautions in collecting the blood for examination than is likely to emanate from the analysis.

Bernard has furnished in the ' Comptes Rendus,'[*] and in his recently published work on Diabetes,[†] a number of results to show the comparative state of arterial and venous blood. There is a striking want of uniformity in the difference observable in the several experiments. Upon one occasion, I observe, he found identically the same quantity of sugar in the blood of the jugular vein as in that of the carotid artery. In all the others the arterial blood gives higher figures than the venous, the difference rising to as much as 0·720 per 1000 —that is to say, an amount considerably larger than the whole quantity which, according to my own experience, exists naturally upon an average in the blood of the sheep and bullock, and nearly the quantity in that of the dog.

I have given the most careful attention to this subject, and performed a large number of experiments bearing upon it. Unless the blood is collected from the artery and vein *at the same moment* the experiment is perfectly useless, and worse than this, for it may be the source of a fallacious inference. Not only must the respective specimens of blood be collected at the same moment, but the animal must have been in a natural and tranquil state just previously ; for experience shows that under almost any deviation from the natural condition, the contents of the circulatory system—and necessarily those of

* Tome lxxxiii, No. 6, p. 373.
† Leçons sur le Diabète. Paris, 1877.

the arteries before those of the corresponding veins
—become charged, to a larger or smaller extent,
with an increased quantity of sugar.

It might be thought that with an increased
quantity in arterial blood the venous blood would
be similarly influenced, and that as long as the
collection is made at the same moment a difference
on the side of excess in the arterial blood might be
taken as evidence of a corresponding destruction in
the systemic capillaries. But such is by no means
to be accepted as conclusively the case, for if from
any cause there has been an increased flow of sugar
into the circulation, when the blood thus charged
reaches the systemic capillaries, the sugar from its
diffusibility will escape by osmosis and give rise
to an appearance of loss by destruction which may
not in reality have occurred.

Observation has taught me the necessity of giving
the most scrupulous care to the manner in which
the blood is collected for analysis to avoid being
exposed to fallacy. I may here cite an example
which shows how speedily an alteration in the condition of the blood from the ingress of sugar occurs.
I attempted to collect arterial and venous blood
separately from the neck of a bullock slaughtered by
the Jewish method—that is, by a sudden incision of
all the soft structures of the neck. At the first
moment following the application of the knife a
single gush of bright-red blood is seen at the
bottom of the incision, but a moment or two later
darker blood may be observed to spurt out in two
distinct streams from the neck. This, from its
colour, I at first took to be blood from the jugulars,

and collected it as such, the bright-red blood issuing with the gush immediately following the incision being put down as the corresponding arterial. Weighed quantities of sulphate of soda had been taken to the slaughter-house, so that the operation for analysis might be commenced before coagulation occurred. Observations were conducted upon the blood of two animals, and these are the results that were obtained :—

Sugar per 1000 parts.

	First portion of blood. Mean		Second portion of blood. Mean.	
Obs. 1 ...	(a) 0·544 (b) 0·555	$\Big\}$ 0·549	... (a) 0·581 ... (b) 0·585	$\Big\}$ 0·583
„ 2 ...	(a) 0·629 (b) 0·620	$\Big\}$ 0·624	... (a) 0·673 ... (b) 0·686	$\Big\}$ 0·679

Thus, in each case the blood supposed to be venous contained a larger amount, to some extent, of sugar than the arterial. I was puzzled to account for this phenomenon, and it was only on a subsequent occasion, when I specially examined the sources of the blood, that I found I had been mistaken, and that the specimens in reality all consisted of arterial blood. The results, therefore, represented the difference producible by the lapse of what appeared an insignificant amount of time, and this as the effect of the simple occurrence of hæmorrhage. In the use of the poleaxe for slaughtering there is a further complication. The injury inflicted upon the brain tells very rapidly in determining a marked influx of sugar into the blood, and affords an additional chance of exposure to fallacy.

Permit me to pause for a moment and direct

attention to the closeness of the results yielded by
the counterpart analyses of the respective specimens
just considered. In no case did the difference reach
beyond 13 milligrammes (about one-fifth of a grain)
upon the amount of sugar contained in 1000 grammes
(upwards of 15,000 grains, or two pounds) of blood,
and in one case the difference only amounted to four
milligrammes. Nothing can testify more strongly
to the delicacy of which the process of analysis is
susceptible.

The considerations to which I have adverted show
how indispensable it is that strict precautions should
be observed in the process of collection, to obtain a
representation of the natural comparative state of
arterial and venous blood. My own experience
irresistibly leads me to the conclusion that these
precautions have not been sufficiently observed by
Bernard, and it is in this way that I explain the
widely different comparative results that he has
obtained. Indeed, in the description of one of his
experiments ('Leçons sur le Diabète;' Paris, 1877,
pp. 229-30) his language leaves it to be directly
inferred that he first operated for the collection of
venous blood, and then for arterial. For instance,
he says that in a moderate-sized dog he exposed the
right crural vein, and inserted a cannula into it and
withdrew 25 grammes of blood. Next follows the
statement that he immediately opened the artery of
the same side, introduced a cannula and allowed 25
grammes of blood to escape. His analysis gave
0·730 per 1000 of sugar for the venous, and
1·450 per 1000 for the arterial blood, representing
a difference of 0·720 between the two, or half the

amount of sugar found in the arterial blood. In
another place (pp. 218-19) he describes an experiment where a first collection from an artery and
vein was made, and a marked difference observed
in the analytical results, whilst a second collection, a
short time afterwards, furnished specimens presenting an identity in the amount of sugar. The particulars of the experiment are these. He placed a
cannula in the left carotid artery of a dog, and
another in the right jugular vein, and withdrew
simultaneously, it is stated, thirty-six grammes of
venous and thirty-seven grammes of arterial blood.
The specimens were submitted to analysis, and the
arterial blood found to contain 1·480 parts per 1000
of sugar, and the venous 1·250 parts per 1000. He
then repeated the collection, withdrawing simultaneously thirty-two grammes of venous blood and
fifty-two of arterial. His analysis now gave 1·560
per 1000 of sugar for the venous, and identically
the same figures for the arterial. Here, he remarks,
the preceding abstraction had modified the amount
of sugar and effaced the previous notable difference
found in the two kinds of blood. The interpretation which my own experience would suggest is that
the conditions of the second collection some time
after the operative part of the experiment had been
performed would be more favourable for supplying correct information regarding the comparative
state of arterial and venous blood. I would also
remark that everything depends upon the exactness
of the proceeding comprehended under the term
" simultaneous " applied to the collection of the
blood. So short a time suffices to produce a

fallacious result that the fallacy may be incurred notwithstanding, ordinarily speaking, it might be said that a simultaneous collection had been made. The observations I have related upon the blood collected from the bullock after the Jewish method of slaughtering exemplify this, and other evidence has been afforded to me of a like nature. It may be inferred also that Bernard has not attached the importance to an *absolutely* simultaneous collection that I have learnt the necessity of doing, or he would scarcely have put forward, without any remark, amongst his experiments, the instance I have cited, where, according to the account given, the collection of blood from the artery followed the completion of the collection from the vein. It was in this instance that the large difference I have commented upon between the arterial and venous blood was noticed—a difference comprised between 1·450 per 1000 (arterial) and 0·730 per 1000 (venous).

Having said thus much about the sources of fallacy to which the experimentalist is exposed in conducting observations to determine the natural comparative states of arterial and venous blood in relation to sugar, I will direct attention to my own results which have been obtained under the exercise of all the care that could be bestowed both upon the process of collection and the steps of the analysis.

In a first set of observations the blood was obtained instantaneously after death. The animal was pithed, and instantly afterwards a scalpel was drawn across the artery and vein determined upon, without any attempt at isolation. The operation was too quickly performed to allow time for an alteration from the natural living condition to occur, and the specimens of blood were collected quite simultaneously. In Observation 1 the blood was obtained from the jugular vein on the one side of the neck, and the carotid artery on the other. In Observations 2, 3, and 4, the crural was selected as the artery. The jugular was still kept to as the vein. The blood throughout the arterial system being the same in character, it is of course quite unnecessary that corresponding vessels should be selected. Any artery that is most conveniently situated for the experiment may be taken.

Amount of sugar found in arterial and venous blood collected simultaneously from the dog instantly after death.

Sugar per 1000 parts.

	Arterial.	Mean.		Venous.	Mean.
Obs. 1 ...	(a) 0·938 (b) 0·915	0·926	...	(a) 0·904 (b) 0·897	0·900
„ 2 ...	(a) 0·799 (b) 0·791	0·795	...	(a) 0·793 (b) 0·791	0·792
„ 3 ...	(a) 0·849 (b) 0·847	0·848	...	(a) 0·847 (b) 0·854	0·850
„ 4 ...	(a) 0·812 (b) 0·830	0·821	...	(a) 0·797 (b) 0·798	0·797

With reference to the last observation, it may be mentioned that my laboratory book contains a note stating that a little difficulty was experienced in collecting the arterial blood. It flowed somewhat slowly, and the last portion of it was dark in colour. In this condition it would doubtless have just commenced to be influenced by the *post-mortem* influx of sugar from the liver. Notwithstanding this circumstance, I have considered it advisable to include the observation in the list.

In a second set of observations to be now referred to the blood was collected during life. Chloroform in some of the experiments and ether in the others were first administered for the performance of the operation of exposure of the vessels. In each case the carotid artery of the one side and the jugular vein of the other were the vessels taken. After being cut down upon, and properly isolated from the surrounding tissues, a ligature was passed underneath, and tied so as to form a loose loop. In this way, the loops being left of sufficient length to reach the surface, the vessels could be pulled forward without exciting any pain or disturbance. A period of an hour and a half or two hours was allowed to elapse after this operative procedure, for the animal to recover from the influence of the anæsthetic. At the end of this time, without the use of any forcible restraint, and whilst the animal was seated quietly on a table, the vessels were drawn forward, and blood collected simultaneously from each. The process of analysis was then at once commenced, and here is a record of the results obtained in seven observations thus conducted.

Amount of sugar found in arterial and venous blood collected simultaneously from the dog during life.

Sugar per 1000 parts.

		Arterial.			Mean.		Venous.			Mean.
Obs. 1	...	(a)	0·806	}	0·811	...	(a)	0·808	}	0·798
		(b)	0·817				(b)	0·788		
„ 2	...	(a)	0·873	}	0·863	...	(a)	0·896	}	0·879
		(b)	0·854				(b)	0·863		
„ 3	...	(a)	0·918	}	0·933	...	(a)	0·918	}	0·916
		(b)	0·948				(b)	0·914		
„ 4	...	(a)	0·870	}	0·884	...	(a)	0·859	}	0·866
		(b)	0·899				(b)	0·873		
„ 5	...	(a)	1·079	}	1·080	...	(a)	1·102	}	1·099
		(b)	1·081				(b)	1·096		
„ 6	...	(a)	1·231	}	1·231			1·240		
		(b)	1·232							
„ 7	...	(a)	1·155	}	1·162	...	(a)	1·180	}	1·183
		(b)	1·170				(b)	1·187		

Average amount of sugar per 1000 parts of blood deducible from the foregoing eleven observations.

Arterial.			Venous.
0·941	...	.·.	0·938

Excess on the side of arterial blood, 0·003.

Such, then—viz., three in the third decimal place —is the amount of average difference given by the observations I have adduced on the comparative state of arterial and venous blood in relation to sugar. On casting the eye through the list, it may be noticed that the corresponding results are throughout closely conformable. In some the excess is on the side of the arterial; in others, on that of the venous; and it is to be remarked that what difference is perceptible scarcely amounts to more than might be legitimately considered as falling within the limits of variation arising from the analysis, for it is not pretended that *absolute* precision admits of being attainable. It must be further remembered that, the results being represented in amount of sugar per 1000 grammes of blood, and the analyses having been conducted upon about 20 grammes, whatever slight error may have occurred becomes multiplied fifty times. A difference, for instance, of $\frac{1}{10}$ of a milligramme ($\frac{1}{650}$ of a grain) in the actual analysis would appear as five milligrammes in the result.

The performance of double analyses gives a trustworthy character to the evidence that it would not otherwise possess, and the closeness noticeable in the counterpart results speaks strongly in favour of the delicacy of the analytical process. In one instance it happens that only a single set of figures is seen to be given, an accidental loss of some of the precipitated suboxide having occurred, and vitiated the counterpart result.

It may be noticed that the amount of sugar found in the blood is higher in some of the observations than what appears to exist under ordinary circumstances. This arises from the effect of the anæsthetic. If the blood be withdrawn whilst the animal is under the influence of the anæsthetic, upwards of two parts per 1000 may be found, and the results included in the list show that even after the lapse of an hour and a half or two hours the sugar may not have fallen to its standard amount.

I have now dealt with the physiological considerations which stand, it may be said, at the foundation of the pathology of diabetes. One cardinal point is that my quantitative analyses show, as I maintained before I was in a position satisfactorily to supply actual quantitative expressions, that there is only a small amount of sugar naturally existing in the blood. A second point is that, corresponding with this small amount of sugar in the blood, there is a certain amount also, something under 0·5 per 1000, present in the urine. A third point comprises the fact that if any difference exist in the amount of sugar belonging to arterial and venous blood it may be spoken of as only of an insignificant nature.

It is hardly possible to overrate the importance of the last consideration. In proportion to the ingress of sugar into the circulatory system, whether from the food or the liver, there must be removal by destruction or otherwise to obviate increasing accumulation. This follows as a necessary conclusion, and put in converse terms the proposition stands, that in proportion to the limitation in the capacity of destruction, so must be the limitation in the supply to the circulation to prevent the occurrence of accumulation. Now, experimental and other evidence shows that corresponding with the amount of sugar existing in the circulation there is elimination with the urine, and therefore whatever accumulation occurs becomes revealed by the condition of the urine. Without a marked capacity for destruction, and in the absence of a marked amount of sugar in the urine, the conclusion is inevitable that the entrance of sugar into the circulation cannot ensue to any marked extent.

Under the presumption that a negative behaviour with the ordinary mode of testing is indicative of an absence of sugar from the urine, Bernard formerly asserted that the limit to the capacity of destruction in the system could be expressed. If, for instance, half a gramme of grape sugar be injected into the jugular vein of a rabbit, there is no reaction given by the urine. If one gramme, on the other hand, be employed, evidence is perceptible of the presence of sugar. Upon these data, about one gramme was assigned as the limit to the quantity of sugar that could be destroyed in the system of a good-sized rabbit. The fallacious character of the ground-work for such a conclusion is now obvious from what has been said about the actual condition of healthy urine, but before our information was as definite as it may be considered to stand now, I pointed out that the observation of Bernard need not necessarily receive the interpretation that was given to it, for it must be remembered that when sugar is introduced into the circulation it will be distributed throughout the system, and by virtue of its diffusibility escape from the blood-vessels into the tissues, a portion only at the time finding its way to the kidney to appear in the urine. When introduced beyond a certain limit a sufficient amount reaches the urine to be easily recognisable, but below the limit the urine remains insufficiently charged to give a distinct reaction under ordinary testing. Such I consider is the true interpretation, and that no evidence of destruction having occurred is afforded.

In a former part of these lectures, when referring to the presence of sugar in healthy urine, I said that I should again have occasion to touch upon this point at a later period. The fact that healthy urine contains a minute amount of sugar is one of great significance. It shows that even with the small quantity of this principle naturally existing in the blood an escape occurs with the urine. Not only this, but all other evidence is to the effect that according to the condition of the blood in relation to sugar so is that of the urine. The urine, in fact, in this respect, partakes of the character, and thus furnishes an indication, of the state of the blood. No matter from what source the sugar reaches the general circulation, it is conveyed in part to the kidney, and, in proportion to its amount escapes with the urine. As I have before stated, it is now contended, under the glycogenic doctrine, that destruction corresponding with ingress occurs in the systemic capillaries. This view is inconsistent with the exigencies of the case. If destruction occurred in the lungs, the course of events would harmonise with the requirements, as the seat of destruction would be situated between the liver—or the point of ingress—and the kidney, and the flow to the kidney would be obviated in proportion to the destruction. With destruction, on the other hand, in the systemic capillaries, whatever amount of sugar passes with the blood to this part of the circulatory systém must also be conveyed in like proportion to the kidney for elimination with the urine to occur. The position would be, that sugar on its way to the seat of destruction would be in part diverted to the kidney, and thence give rise to escape from the system.

Upon the strength of his quantitative determinations latterly conducted, Bernard speaks of a tolerating capacity of the blood for sugar which prevents its passing off with the urine. He has given to this tolerating capacity a definite or sharp line of limitation ; his statement being that if the blood of the dog contain 2·40 per 1000, sugar does not appear in the urine, whereas with 2·60 per 1000 it does. The mean, or 2·50 per 1000, he puts forward a sexpressing the limit of toleration in this animal.

Looked at, to begin with, from the point of view of feasibility, I think it may be said that this assertion bears upon the face of it a strong impress of improbability. That a hard line should exist between the escape and non-escape of a diffusible substance like sugar from the blood certainly seems extraordinary. But we have the fact that with the small quantity of sugar naturally present in the blood a certain amount of escape occurs with the urine. Observation also in a direct manner shows that in diabetes there may be voidance of a notable amount of sugar with a proportion in the blood considerably short of the alleged limit of toleration. Here are results illustrating the relation existing between the blood and urine of diabetics under different states as regards degree of severity of the disease. Nothing can be required more clearly to show that the character of the urine is expressive of that of the blood. This applies equally throughout, and therefore stands good for small quantities as well as large. In proportion to the ingress of sugar into the circulation, so is the escape with the urine. For the urine to present the character belonging to health there must only be the small proportion of sugar in the blood which I have represented as naturally to be found.

The table needs but little explanation. The blood examined consisted of blood removed by cupping. The particulars regarding the urine refer to the twenty-four hours' period within which the cupping was performed. The first three patients suffered from severe forms of the disease. The last case was one of a mild character, the patient throughout consuming a moderate quantity of ordinary bread, but otherwise following a restricted diet.

COMPARATIVE STATE OF BLOOD AND URINE IN DIABETES.

	URINE.					BLOOD.
	Quantity per 24 hours.	Specific gravity.	Sugar per 1000 parts.	Sugar, grains per fl. oz.	Sugar per 24 hours.	Sugar per 1000 parts, mean of two analyses.
CASE 1. Jan. 5th.—Mixed diet.	232 fl. oz. (6608 c.c.)	1040	109·91	50·00	11,600 grains (751·6 grammes).	5·763
CASE 2. Jan. 8th.—Mixed diet.	228 fl. oz. (6474 c.c.)	1041	94·08	42·85	9769 grains (633·0 grammes).	5·545
Jan. 28th.—Restricted diet.	120 fl. oz. (3407 c.c.)	1081	61·34	27·69	3322 grains (245·2 grammes).	2·625
CASE 3. June 8th.—Mixed diet.	207 fl. oz. (5878 c.c.)	1036	93·39	42·33	8762 grains (567·7 grammes).	4·970
July 20th.—Restricted diet.	87 fl. oz. (2470 c.c.)	1033	45·49	20·55	1787 grains (115·8 grammes).	2·789
CASE 4. March 9th.—Partially restricted diet.	60 fl. oz. (1704 c.c.)	1036	48·11	21·81	1308 grains (84·1 grammes).	1·848
June 28th.—Partially restricted diet.	30 fl. oz. (852 c.c.)	1034	31·76	14·40	431 grains (27·9 grammes).	1·543

The results exhibited in the table I consider are extremely instructive. They show that the condition of the urine closely follows that of the blood. Even with 1·543 (figures much below those expressive of Bernard's tolerating capacity) per 1000 of sugar in the blood, as in the last example in our table, the urine is seen to have contained 31·76 per 1000, or 14·40 grains to the fluid ounce—an amount which may be spoken as of a decidedly notable nature, taken even in relation to diabetes. The reason that healthy urine does not present the character of that of diabetes arises from the amount of sugar allowed to reach the circulation being as small as it is. Did the condition which Bernard has contended for exist, the urine of all would be strongly diabetic. The facts before us lead irresistibly to this conclusion.

I strenuously maintain that instead of the liver being essentially a sugar-forming, it is a sugar-assimilating organ. Its great function in relation to sugar is to prevent this principle reaching the circulation to any material extent. I do not deny, and never did deny, that sugar in minute quantity reaches the blood. The facility with which the amyloid substance in the liver is transformed into sugar renders it surprising that more is not permitted to enter ; but it is one of the conditions of health that such should be the state of things existing. Under a variety of deviations from the healthy state, on the other hand, sugar in more or less notable quantity does reach the circulation, and, as a consequence, finds its way correspondingly into the urine.

From what has been said, it will be seen that the

6

amount of sugar eliminated is expressive of the amount entering the circulation. Whatever proportion is conveyed to the peripheral capillaries must also be conveyed to the kidneys. Normally this proportion is so small that the urine examined in the ordinary manner does not show a decidedly saccharine behaviour. If sugar in larger amount reach the circulation, it becomes revealed by a proportionate presence in the urine, and it is just this latter state which constitutes a deviation from, instead of a representation of, the natural condition.

I have always understood the glycogenic function to mean an operation associated with the application of sugar to heat, or some other kind of force-production. The appropriation of sugar in some way or other to this purpose must constitute a large and important operation of life from the position held by the carbo-hydrate group of principles in vegetable food. The production of sugar by the liver has been regarded as complementary to the introduction of sugar with the food, and as placing the carnivorous animal in an analogous position to the herbivorous in relation to the principle in question. Such are the terms under which the glycogenic doctrine was propounded, and it is maintained that sugar is formed, and discharged into the current of the blood for the purpose of being conveyed to the peripheral part of the circulatory system to undergo utilisation by application to force-production, in the same manner as has been considered to be the case with sugar supplied from without.

Now, it is just this which I object to. I have

shown upon what physiological grounds my objection is based, and I have further to say that the
doctrine is incompatible with the phenomena of
health considered in relation to those of diabetes.
For instance, let it be supposed that we have before
us two persons, one in a state of health, and the
other suffering from an ordinary form of diabetes,
and that we supply them with the same allowance of
mixed food *per diem.* The healthy person elimi-
nates no material amount of sugar with his urine,
whereas the diabetic voids it to the extent, it may
be, of a pound (453·6 grammes), or more, per diem.
On looking at the state of the blood, it is found that
in the healthy person sugar is only present in
minute proportion, whilst in the diabetic it exists to
a notable extent. So far, I have only given a
simple expression of facts, and let us see what
argument is deducible from them. In the diabetic,
sugar reaches the general circulation, in part
derived directly from the amylaceous and saccharine
principles of the food, and in part, it may be said,
from the action of the liver—a true glycogenic
action—upon nitrogenous matter ; for if lean meat
alone be consumed, sugar is still found to be
voided, although in greatly diminished quantity.
The effect of the sugar from ingestion and the sugar
from formation reaching the circulation is to lead to
the production of more or less strongly saccharine
urine. The saccharine condition of the urine stands
in proportion to the amount of sugar allowed to
reach the general circulation. Such is the condition
existing in diabetes. In the healthy state, on the
other hand, the urine escapes being similarly

charged with sugar, and the inference is to be drawn that sugar does not similarly reach the general circulation. The phenomena alleged to occur under the glycogenic doctrine are precisely those which give rise to diabetes. If, in fact, the sugar derived from ingestion, and that presumed to be formed by the liver, entered the general circulation, we should all be in the position of the diabetic. The question is one which bears upon the destruction of an important part of our utilisable material, and the fact that this material is not eliminated with the urine in health as it is in diabetes is proof that it passes in a different direction within the system under normal conditions from that assumed by the glycogenic theory. In other words, there must be some other method of appropriation besides the passage into the blood for destruction in the peripheral capillaries. I take it, as there are grounds for doing, that the fault in diabetes essentially lies with the passage of sugar into the general circulation in opposition to what ought to occur.

It is a minimum amount of sugar in the blood which is the character of health, and a larger amount which is associated with unnatural states. To maintain the minimum amount the liver exercises a sugar-detaining and sugar-assimilating function, and this prevents us from being diabetic. It has been asserted by Bernard, and my own observations confirm the assertion, that the blood of the herbivora contains no more sugar than that of the carnivora. Indeed, according to the results I have furnished in a former part of these lectures, the blood of the dog contains a higher proportion of sugar than that of the sheep or bullock. Now in the herbivora, besides the sugar derivable from an internal source, we have that to deal with proceeding from ingestion; and if the liver did not stop, as I contend it does, the passage of sugar into the general circulation, there ought to exist a marked difference adversely to what is noticed. The sugar which actually escapes from the liver is so small in amount that I cannot conceive it to have anything to do with the large operation of appropriation of this principle to force-production that must occur as a part of the general phenomena of life.

Expressed in precise language, I would say that the liver is essentially a sugar-assimilating instead of a sugar-forming organ; and that when its assimilative action is properly exerted, so little sugar is allowed to pass into the general circulation that the quantity existing in arterial blood is insufficient for rendering the urine more appreciably saccharine than is observed in the healthy state : but that when its assimilative action is not properly exerted, sugar is allowed to pass, and in proportion as it does so the urine acquires a more or less marked saccharine character.

The question before us is one which resolves itself into the utilisation or non-utilisation of a certain kind—viz., the carbo-hydrate kind—of material. We know that in diabetes this material reaches the blood as sugar, and thence escapes unutilised with the urine. If in health it passed in a similar manner into the blood, to obviate its escape it must undergo appropriation in the circulatory system, and the difference between health and diabetes would then resolve itself into a difference in the capacity of appropriation within the circulatory system. This view, however, is not adopted by the advocates of the glycogenic doctrine, and the evidence afforded by observation is against it. We are therefore driven to the conclusion that the material in question must have a different destination under normal circumstances, and I contend that in reality the liver ought to detain and assimilate it instead of allowing it to pass as sugar into the circulatory system, and that it is by this action we escape being diabetic. Put in another way, it may be said that the sugar escaping in diabetes is the representation of the carbo-hydrate material which ought to be utilised in the system, but which, finding its way into the general circulation as sugar, is thence eliminated as it is. To escape being thus eliminated, it is obvious that the carbo-hydrate material in question must pursue some other path towards utilisation than passage into the general circulation as sugar, and this is tantamount to saying that under normal conditions there must exist a capacity of appropriating carbo-hydrate material in this other direction equal to the disposal of that which is represented by the sugar escaping in a full-marked case of diabetes, and this embraces, if not all, certainly in the main, all the carbo-hydrate material which the system has to deal with. Not only, therefore, must there be this other power of disposal, but such power must be exercised, instead of the material being allowed to pass into the general system as sugar as is represented under the glycogenic doctrine, to avert the manifestation of diabetes.

I have previously discussed what Bernard has said about the appropriation or destruction of sugar in the circulatory system, and have shown that my own analytical observations give no evidence of any significant amount of destruction occurring. It may be assumed—as, indeed, the aggregate of my results indicated—that a minute disappearance takes place. With an organic principle like sugar, susceptible of undergoing change such as occurs in the lactic acid fermentation, it is scarcely to be expected that it would remain uninfluenced by the molecular actions going on around it; but that nothing of a nature commensurate with the large disposal of it, that there must be as an alimentary principle, is to be detected is, I consider, proved by the evidence that has been already adduced. If the destruction of sugar occurred within the circulatory system of the herbivorous animal to an extent sufficient to account for the disposal of that which is derived from ingestion as well as from alleged formation, what, it may be asked, ought to be the condition in the carnivorous animal which does not receive the supply of sugar from without? Surely, if the destruction is carried out with sufficient fullness to keep down the sugar in the system of the herbivorous animal, it ought to be equal to the entire removal of the sugar contained in the arterial blood during its passage to the veins in the case of the carnivorous animal. In reality, the sugar that has to be dealt with in the blood shows itself to be as much beyond the capacity of destruction in the peripheral vessels in the carnivorous as it is in the herbivorous animal. No material difference, in fact, is to be traced in the state of things in the two groups of animals. In both cases alike the sugar contained in arterial blood reaches the veins without undergoing any material diminution, and thence must needs circulate through the system over and over again. Whatever destruction occurs can only fulfil a subsidiary office. Admitting that the operation is to some extent carried on, it is altogether wanting in the magnitude required to be associated with the appropriation of sugar as a force-producing agent in the body.

My own version of what occurs is this. We know that in a large number of animals the food is of such a nature as to supply sugar in notable quantity for absorption from the alimentary canal. A portion of this may reach the thoracic duct through the absorbents, and thence be conveyed to the general circulation, accounting in part for what sugar is there encountered. The main channel, however, for the passage of sugar from the alimentary canal appears to be the bloodvessels, and transmission into them is permitted by the property of diffusibility which this agent possesses. Absorbed into the portal system, it is conveyed to the liver, where it becomes if not entirely, certainly almost entirely, checked in its onward progress, and prevented from entering the general circulation. I have shown, by observations which have been for a long time made known, and which I need therefore here only allude to, that sugar leads to an increased accumulation of amyloid substance in the liver. The evidence adducible points to the occurrence of a direct formation of amyloid substance from absorbed sugar. As the sugar is passing through the capillaries of the organ it becomes picked out by its cells and converted into the principle I have mentioned. Nothing could be more favourable than the conditions that exist for the occurrence of this operation. Surrounded as the capillary vessels of the lobule are with rows of secreting cells, the blood is brought into the closest proximity with the secreting element, and thus most advantageously placed for selective action to be exerted.

Micro-chemical observation shows that the amyloid substance of the liver is confined to the cells; and, doubtless, immediately the sugar is abstracted from the blood it is transformed by the action of the cell into the principle named, for otherwise the organ would give evidence of the presence of sugar in a manner that it is not found to do. Moreover, sugar is a diffusible substance, and would not therefore be retained. Amyloid substance, on the other hand, belongs to the class of colloids, and with its non-diffusibility possesses the properties which physically contribute to its retention in the cells where it presumably undergoes a change which forms one of the links in the series leading up to the final issue—the utilisation of sugar as a force-producing agent in the system.

As regards, then, the sugar derived either directly or indirectly from the food and absorbed from the alimentary canal, my proposition is that under normal circumstances it is stopped by the selective or secreting action of the cells of the liver, and in these transformed into amyloid substance. When not thus stopped it reaches the general circulation, and as a result gives rise to a saccharine impregnation of the urine, standing in proportion to the amount of sugar absorbed from the alimentary canal. This is just the condition that exists in diabetes, and it is well known that in this disease the eliminated sugar stands in relation to the amount of sugar or sugar-forming material ingested.

But it is not only from the sugar derived from the food that the amyloid substance of which I have spoken takes origin. I do not know of any evidence to show that it is formed from fat. Nitrogenous matter, however, undoubtedly is a source for it. This is proved by its existence in well-marked quantity in the liver of the animal-feeder kept upon lean meat. In diabetes also, where a severe form of the complaint exists, sugar is voided upon a strictly animal diet, and such sugar may be put down as taking origin from the abnormal descent of the amyloid substance derived from nitrogenous matter. It is known that ingested nitrogenous matter leads to the production and elimination of urea, and it is believed that the production occurs in the liver. A splitting up of the nitrogenous compound ensues, and its nitrogen, appropriating a certain amount of its carbon, hydrogen, and oxygen, to form urea, leaves a complementary portion of carbon, hydrogen, and oxygen from which it may be assumed amyloid substance is generated. If we take the per-centage composition of albumen and give to the nitrogen the amounts of the other elements required to form urea, we see the amounts of carbon, hydrogen, and oxygen that may be applied to the purpose named. This table affords a view of how the matter stands :—

Urea and Complementary Residue derivable from Albuminous Matter.

	Albumen.	Urea.	Residue.
Carbon	53·50	6·64	46·86
Hydrogen	7·00	2·21	4·79
Nitrogen	15·50	15·50	—
Oxygen	22·00	8·85	13·15
Sulphur	1·60	—	1·60
Phosphorus	0·40	—	0·40
	100·00	33·20	66·80

From this tabular representation it is seen that in the presumed splitting up of the albuminous molecule one-third passes into urea and two-thirds into complementary residue—that is, the relation between the urea and complementary residue is as 1 to 2. Now, some observations by Dr. Sydney Ringer on the relation between eliminated urea and sugar in diabetes, that were published in the 'Medico-Chirurgical Transactions' for 1860, present a conformity which is certainly striking, if nothing more. The observations show that during fasting and under a diet of animal food—when, in other words, there is no sugar from ingestion to influence the state of the urine—the eliminated urea and sugar stand in the ratio of 1 to 2·2 As might be expected, the figures representative of the eliminated products are much higher under the animal diet than during fasting, but notwithstanding this difference, the same ratio was maintained. It is right to state that this correspondence I have brought into view must not be taken for more than it is worth, inasmuch as the complementary residue referred to does not express the composition of sugar or of one of the allied carbo-hydrates, the carbon being out of proportion to the hydrogen and oxygen, and particularly so to the latter.

Whatever may be the precise manner, however, in which the elements of nitrogenous matter become re-arranged, it may be said that there are grounds, on the one hand, which have led observers to entertain the view that it is in the liver the production of urea occurs, and practically, on the other hand, it can, as I have mentioned, be shown that from nitrogenous matter amyloid substance is here formed. It is the latter point which essentially concerns us with reference to diabetes. The liver not only arrests the passage of sugar absorbed from the alimentary canal, and effects its transformation into amyloid substance, but also forms this substance from nitrogenous matter. This may be regarded as constituting the first step of an assimilative action exerted by the organ; and, if we are not able at present to follow the process on, and state what next occurs, we can, I consider, say upon the evidence I have advanced that the amyloid substance is not physiologically destined to undergo conversion into sugar, and pass as such into the general circulation. The glycogenic doctrine, it is true, implies the occurrence of such an event. Transformation into sugar is contended for, and Bernard speaks of this formation of sugar as the beginning of a series of phenomena of combustion ultimately resulting in carbonic acid and water.

It is one of the properties of amyloid substance to

possess a strong tendency to pass into sugar under the influence of contact with bodies of the nature of ferments. Various unnatural states, and the condition existing after death, lead to its passage in this direction ; but that such is not its proper destination, certainly not to any material extent, is shown, I maintain, by the several considerations I have advanced. Why under natural circumstances it should be capable of resisting, as it is contained in the hepatic cells, the action of the ferments around, whilst under other circumstances it does not, is a problem which I cannot attempt to solve. It can only be said that such is observed to be the fact ; and with regard to the well-known phenomenon of the coagulation of the blood, it happens that we are situated in a position of equal difficulty. If what is meant by the glycogenic function of the liver applies to that minute amount of sugar (I have always spoken of a minute amount of sugar existing in the blood) which is allowed to reach the circulation—that which, through its being minute, obviates the production of the diabetic state,—I readily concede that such a function exists. But such are not the terms under which the glycogenic doctrine has been advanced, and it is the passage of sugar from the liver as an incident belonging to the large operation of appropriation of this principle to force-production that I contend is incompatible with the evidence derivable from the facts before us. I do not simply say that the liver does not transmit sugar into the circulation for the purpose named, but that it exerts an active part in preventing it from so passing, and it is through this active part we escape being diabetic.

In addition to other considerations, my argument is, that in proportion as sugar reaches the general circulation so it shows itself in the urine; and the fact that sugar does not appear in the urine in a healthy state, as it does in diabetes, is proof that the material involved in the question must be appropriated in some other way instead of being allowed to pass as sugar into the general circulation. If the material, in fact, with which we are concerned— and it includes the carbo-hydrate principles ingested and produced naturally—passed as sugar into the general circulation, the accumulation which occurs in diabetes could only arise from a subsequent faulty process of destruction. There is no alternative to fall back upon. But supposing it to be the natural condition that sugar reached the circulation for subsequent destruction in the system, it would be conveyed to the kidney as well as to the general capillaries, and in proportion to the amount furnished for destruction so would be the amount for elimination with the urine. The state in which this secretion is normally found shows that this hypothesis cannot hold good.

I consider that the great service which has been rendered by Bernard in relation to this matter is the discovery of amyloid substance. Without this discovery we could not have attained the position we now stand in. Science is, therefore, deeply indebted to him for the perseveringly and skilfully conducted researches which placed us in possession of knowledge the importance of which it is almost impossible to overate. I protest, however, against the application of the term " glycogen " to this

material, from the conviction that looked at physiologically it bears an impress of error. Moreover,
in its terminal character it stands at variance with
the chemical nomenclature adopted for the bodies to
which it is allied. I do not say that "amyloid
substance" possesses any other claim than that of
signifying what the body is, without involving any
physiological error. I think it would be worth
while for it to be renamed, and "zoamylin," which
I remember seeing suggested by some one many
years back, would be scientifically the most appro-
priate designation that could be employed. There
is danger, however, in now adopting this name, that
some confusion might be created, and it appears to
me that it would be a fitting tribute to the memory
of Bernard, as the discoverer of the substance, to
call it after him "Bernardin," which would fall in
with the ordinary terminal construction adopted,
and would be of itself at once suggestive of the
body referred to. I throw this out for the consideration of chemists and others, and should the
suggestion meet with favour, I would myself hereafter employ the term.

I now come to the question why sugar should
reach the circulation as it does in diabetes in opposition to what occurs under normal circumstances.
Instead of the sugar absorbed from the alimentary
canal being arrested by the liver, and the amyloid
substance being restrained from passing to any
material extent into sugar, the phenomena of diabetes show that in this disease the sugar of ingestion
is allowed to reach the general circulation, and the
amyloid substance to be converted into and escape

as sugar. Whether the sugar from ingestion passes directly through the liver, or whether it is first transformed into amyloid substance, and this re-converted into sugar, I cannot say ; but the sugar eliminated with the urine, standing in relation as it does to the sugar derivable from ingestion, renders it evident that either one or other must occur. Plainly represented, the problem to be solved may be thus expressed : Why is it that sugar passes into the general circulation in diabetes, instead of being prevented doing so to more than what may be spoken of as an insignificant extent conformably with the healthy state ? Let us see what solution of this problem can be offered.

Many years ago I noticed, after ligaturing the portal vein, and thus allowing only the blood of the hepatic artery to reach the liver, that the contents of the circulatory system became highly charged with sugar. I was struck with the phenomenon, seeing that the amount of blood passing through the liver was so much diminished, and commented upon it as furnishing a link of evidence opposed to the glycogenic theory. I did not find that the urine became saccharine, but this was not to be wondered at. The operation leads to such an accumulation of blood in the portal system, and therefore deviation from the general circulation, that the supply to the kidney may well be expected to be insufficient to permit the secretion of urine, certainly to any material extent, to occur. Bernard, however, has spoken of having observed the presence of sugar in the urine after ligature of the portal vein. So far his experience tallies with the effect I have de-

scribed as produced upon the blood, but I imagine
that his ligature must have been placed low enough
for some of the vessels to have been left free; for
he speaks of his animals as having recovered from
the operation, whilst in my own experiments the
state induced by the blockage of the circulation
appeared to be incompatible with anything more
than a few hours' existence.

Struck with this effect of allowing only the
arterial supply to reach the liver, I afterwards tried
to connect the portal with the right renal vein in
such a manner that the circulation might be maintained with the portal stream diverted from the
liver.   Success, however, was rendered hopeless by
the coagulation which occurred in the connecting
tube.

Later on it occurred to me to ascertain what
would be the result of injecting defibrinated arterial
blood into one of the veins of the portal system.
The idea was a fortunate one, for the issue of the
experiment appears to me to supply the key to the
elucidation of the cause of the passage of sugar from
the liver into the circulation in diabetes.

My results were communicated to the Royal
Society, and published in the 'Proceedings' for
June and November, 1875.   They show that by
introducing defibrinated arterial blood into the
portal system strongly marked glycosuria is quickly
induced.   When I started upon the investigation I
was quite unprepared for the evidence that became
revealed, and before committing myself to a conclusion, I established by the negative effect, observable
under the employment of defibrinated venous blood,

7

that the glycosuria was due to the influence of the oxygenated blood, and not to any other part of the operation.

I may incidentally refer here to a piece of information I came across in performing these experiments, which may possibly prove of assistance to those who are engaged in investigations bearing upon the coagulation of the blood and the office of the spleen. My previous experience of the impunity with which defibrinated blood may be injected into one of the general veins—the jugular, for instance—did not lead me to expect that anything would occur to impede the accomplishment of the object I had in view. To my concern, however, I found that the introduction of the blood into the portal system was apt to be followed by a complete stoppage of the portal circulation, from the occurrence of coagulation in the trunk of the portal vein and its ramifications in the liver. This of course completely frustrated the experiment, and in seeking to extricate myself from the difficulty I tried the effect of extirpating the spleen before the performance of injection. I was led to do this from noticing that it was in the trunk of the portal vein and in the vessels further on, and not at the immediate seat of injection, that the clotting of blood took place. I found that the effect was to produce what I wanted. No coagulation now occurred. A further effect noticeable was that the blood throughout the system after death was restrained from undergoing coagulation in the ordinary manner. Besides the two fibrin-forming factors—fibrinogen and fibrinoplastin, it appears from recent research that the agency of a fibrin-producing ferment is wanted for inducing coagulation, and if any view were expressed, the conjecture might be hazarded that it may be upon this latter that the extirpation of the spleen in some manner or other tells in determining the result I have alluded to.

Returning from this digression to the subject before us, I desire to lay stress upon the fact that the effect of arterial or oxygenated blood introduced into the portal system is to cause sugar to escape from the liver to a sufficient extent to induce strongly marked glycosuria. In half an hour, for instance, after the completion of the injection, I have noticed as much as fifteen grains of sugar to the fluid ounce in the urine—that is, a proportion of about 33 per 1000.

I regard the knowledge thus acquired as of cardinal importance in relation to diabetes. We see how, by a simple alteration in the character of the blood going to the liver, an altered action occurs in the organ which leads to the escape of sugar and the production of glycosuria. No new agent is called into play. The passage of blood through the vessels of the chylopoietic viscera in such a manner as to reach the portal vein in an imperfectly de-arterialised state supplies all that is required to account for the appearance of sugar that is noticed in diabetes. I shall show further on how the blood-vessels may be concerned in determining an imperfectly de-arterialised condition of the portal blood, and how the nervous system, by its influence upon the blood-vessels, may stand at the foundation of all. What I wish at present to draw attention to is simply the fact that oxygenated blood reaching the liver through the portal vein is perversive of the proper action of the organ and instrumental in producing glycosuria.

Upon discovering that glycosuria was produced by the introduction of oxygenated blood into the portal system, it occurred to me that I might possibly be able to surcharge the blood with oxygen through the medium of the respiration, and obtain a similar result. I first tried the effect of causing an animal to breathe oxygen instead of air, and on two occasions, amongst several experiments upon the dog, succeeded in inducing a strongly-marked saccharine condition of the urine. In one of the instances the amount of sugar noticeable was 10·48 grains to the fluid ounce (about 23 per 1000), and in the other 5·71 (about 13 per 1000).

I take it that the reason the inhalation of oxygen did not in all my experiments lead to glycosuria is to be found in the uncertainty of getting the blood in reality surcharged with oxygen. Physiology teaches us that the respiratory movements are governed by the condition of the blood. There is a kind of self-regulating action in operation produced by the influence of the condition of the blood upon the respiratory nervous centre—the accumulation of carbonic acid increasing, and that of oxygen diminishing the stimulus through which the muscles of respiration are called into play. Hence under the respiration of pure oxygen there may be in reality no material alteration produced, the muscular movements adapting themselves to the increased efficiency of the oxygenating process. Indeed, I noticed in some of my experiments that a state of apnœa—I use this word in its strict sense as signifying cessation of respiratory movement—was temporarily induced; and at first I did not know what it meant, and thought that the animal's life was in jeopardy. In those instances where glycosuria was induced it was specially observed that the respiratory movements were carried on in a deep and excited manner.

In addition to the experiments upon the dog, I conducted a number upon frogs, placing the animals in a sealed receiver, and passing a stream of oxygen through it for the space of three or four hours. In some instances no effect upon the urine was perceptible, but in others a distinctly saccharine behaviour was noticed. The urine of the frog is readily obtained by means of properly applied pressure over the bladder; and, by using a small test tube, the amount procured from one animal was generally sufficient for testing.

Finding that the result in the experiments with the inhalation of oxygen was dependent upon the chance degree of activity with which the respiratory movements were carried on, I tried, later on, the effect of performing artificial respiration. An artificial respiration apparatus was employed, which enabled me to inflate and exhaust the chest at any rate of frequency I desired. By its use I found that the blood could be readily surcharged with oxygen to a sufficient extent to induce a prolonged state of apnœa observable on discontinuing the performance of the operation.

Now, by means of artificial respiration conducted

in this way, and applied to the dog, I succeeded in producing strongly marked glycosuria. At first I employed oxygen, but afterwards found that atmospheric air equally answered the purpose. In my communication to the Royal Society to which I have already referred, it is stated that in one of the instances the analysis of the urine showed the presence of eleven grains of sugar to the fluid ounce, or about 24 parts per 1000.

I may mention that there is one observer, Tieffenbach, who has anticipated me in obtaining glycosuria by artificial respiration. In referring to the production of glycosuria in curarised rabbits subjected to artificial respiration he incidentally states that he once obtained the same result without the previous poisoning.

It is a point of interest that a gas—viz. carbonic oxide—which produces the same physical effect on the blood as oxygen, likewise occasions glycosuria. The bright scarlet colour belonging to arterial blood is also the colour produced by the influence of carbonic oxide; and it has been further shown by Dr. Gamgee that the spectral properties of the compound formed by the colouring matter with carbonic oxide are identical with those of oxidised blood. Carbonic oxide, however, stands in a different position from oxygen, as regards tenacity of union with the hæmoglobin; and whilst assuming that their mode of action may be the same in determining glycosuria, we can discern in this difference an explanation of the difference noticeable in the degree of facility with which the phenomenon is producible, for experiment shows that the inhalation of carbonic oxide affords one of the most speedy means of causing the urine to become strongly charged with sugar. I apprehend in the case of both gases that the glycosuria is due either to the gas itself or the compound formed with hæmoglobin exciting a transformation of the amyloid substance of the liver into sugar. Now, it happens that the combination between oxygen and hæmoglobin (oxy-hæmoglobin) is so feebly maintained that the gas is readily displaced from the corpuscle by the agency of carbonic acid. Hence the facility with which the oxygen is yielded up in the systemic capillary circulation, and the blood reduced from the oxidised to the de-oxidised condition, in which state it is thereby allowed to reach the liver, and, as observation shows, does not act like oxidised blood in leading to the conversion of amyloid substance into sugar. With carbonic oxide, on the other hand, its difficulty of displacement from the corpuscle is known to be a marked feature belonging to it. As the result of its inhalation the blood becomes charged with it, and reddened as it is by oxygen; but, on account of the tenacity with which its combination with hæmoglobin is maintained, it is not similarly yielded up in the systemic capillaries. It is thence allowed to reach the liver, and act in the manner I have suggested in producing glycosuria. The effect of oxygenated blood on the liver shows that if oxygen were held by the corpuscle in the same way as carbonic oxide, it might be fairly assumed that glycosuria would be as readily produced by the one as by the other. It is owing to the facility, it may be considered, with

which oxygen is removed from the blood during its
passage through the systemic capillaries that we
escape being glycosuric under the process of respiration carried on in an ordinary manner in the
atmosphere ; but should the oxygen introduced into
the blood surpass, as it may be made to do by the
experimental methods I have referred to, the capacity of being abstracted to an extent to produce a
natural venous condition of the fluid within the
veins, then, as we have seen, glycosuria is induced.

It appears to me that what we learn from the
action of carbonic oxide affords assistance towards
the comprehension of the subject before us through
the medium of analogy. Oxygen and carbonic acid
form allied compounds with hæmoglobin. The
presence of oxygenated blood and blood impregnated with carbonic oxide in the portal system both
occasion glycosuria. The great difference noticeable in the relation of the two gases to the blood
is, that the one is readily parted with, whilst the
other is strongly held, and it is just this difference
which accounts for ordinary respiration being un-
attended with the effect which follows the inhalation
of even a small quantity of carbonic oxide. It can
be readily shown in the case of carbonic oxide that the
production of glycosuria is not due to an increased
afflux of blood to the liver, through an influence
upon the blood-vessels ; and if such be true for carbonic oxide, there are grounds for believing it to be
equally true for oxygen. I have found, for instance,
after ligature of the hepatic artery, ligature of the
superior mesenteric artery, and even ligature of the
cœliac axis, that the inhalation of carbonic oxide has

been followed by the usual strongly marked gly-cosuria. It may be assumed that it is to the effect upon the blood that the result is attributable, and that blood, either unnaturally charged with oxygen or impregnated with carbonic oxide, acts upon the amyloid substance in such a manner as to lead to its abnormal transformation into sugar.

If we give attention to the positions in which amyloid substance is found in the system it will be perceived that venous blood is favorable, and oxy-genated blood unfavorable, to its accumulation. Let us see to what extent this can be shown to be the case. The circumstance is one which har-monises with the considerations I have just dis-posed of.

There is no organ in the body supplied with venous blood in like manner to the liver; and, in corre-spondence, nowhere does amyloid substance exist to a like extent.

In contrast to the condition of the adult liver, the organ during the first portion of intra-uterine life is free from amyloid substance. It is a curious fact, which was many years ago pointed out by Bernard, that during the first portion of foetal life the liver is free from amyloid substance, and sugar is found in the *liquor amnii* and the fluid of the allantois; whilst during the latter portion, the liver contains amyloid substance, and no sugar is met with in the fluids that have been mentioned. This has always stood as a surprising and inexplicable phenomenon, but with the knowledge now possessed regarding the effect of oxygenated blood it admits of rational explanation. The umbilical vein which

contains arterial blood is connected with the liver in
its passage to the inferior cava, and must mainly
supply the organ at an early period of fœtal life.   It
happens at this period that the liver presents a state
of development greatly out of proportion even to
every other part of the organism.   Text-books on
anatomy tell us that in the human fœtus the organ
at the third or fourth week of embryonic life constitutes one half of the weight of the whole body. Under
such circumstances the supply of venous blood to the
liver from the chylopoietic viscera must therefore be
quite insignificant, and with the oxygenated blood
from the umbilical vein the conditions will be such
as to promote the transformation of amyloid substance into sugar, thereby leading to the absence of
the former principle in the organ, and the presence
of the latter in the *liquor amnii* and the fluid of the
allantois. As fœtal life advances the relation between
the liver and the chylopoietic viscera as regards
development becomes altered.   The former no
longer continues to hold the same prominence over
the latter, and with the increasing growth of these
latter it follows that the amount of portal blood
acquires more and more significance.   Without
appealing to any consideration regarding the precise
anatomical disposition of the umbilical vein in relation to distribution to the liver it may be said to
constitute a necessary consequence of the development of the chylopoietic viscera that the liver must
receive a relatively larger supply of portal blood,
and in proportion as this advance occurs and a pre-
ponderance of portal blood reaches the liver, so will
the conditions be favorable to the accumulation of

amyloid substance instead of the occurrence of its transformation into sugar. Thus, the different states observed stand in harmony with what might be expected under the respective circumstances existing.

Whilst the liver forms the chief seat of amyloid substance in the body of the adult it is not absolutely restricted to this organ. It has also been recognised in the pulmonary and muscular tissues, but observation shows that when present under physiological circumstances in these structures it is only so to a comparatively insignificant extent, and this at a time when a reduced supply of arterial blood prevails. For instance, it is especially in muscles that have existed in a state of rest, and in those of the hybernating animal during the period of hybernation, that it is to be discovered, and it has been observed to disappear under the resumption of activity. The tissue of the lungs has been found to yield it during hybernation, but under other natural states it is absent.

In the case of the solidified lung of pneumonia, however, amyloid substance is to be met with to a notable extent. This I look upon as an exceedingly interesting fact, for it happens that under the circumstances existing, we have a condition as regards blood supply identical, it may be said, with that of the liver. The lungs, like the liver, receive arterial blood through the bronchial arteries, which may be compared to the hepatic artery, and venous blood through the pulmonary artery, which for our illustration may be looked upon as holding the same position as the portal vein. In a natural state, the venous blood loses its character and becomes oxygenated within the organs, thus furnishing a condition which is antagonistic to the accumulation of amyloid substance. In the solidified lung tissue of pneumonia, however, on account of its imperviousness to air, the venous blood of the pulmonary artery will retain its venous character and stand in the same position as the portal blood does to the liver. With blood supplied through the bronchial arteries and the pulmonary artery, and the lung tissue in a state of solidification, a disposition exists which is identical with that belonging to the liver; and under these circumstances it is noticed that amyloid substance accumulates whilst under a natural state of functional activity it does not. Evidence such as is here presented strongly points to its being through its exceptional supply of venous blood that the liver is placed in a position for performing what it does in relation to amyloid substance.

I may refer to one more circumstance bearing on the condition of the blood in relation to the accumulation of amyloid substance. In the fœtus this substance is found in a number of structures from which it is absent in the adult. I would suggest that it is to the partially venous state of the blood distributed to the tissues of the fœtus in contradistinction to the state existing in the adult that the difference noticeable is due.

From all these considerations, which tally with and strengthen each other, we may generalise and say that amyloid substance is a body which tends to accumulate in certain animal structures under

the existence of a limited supply of oxygen, and
that it is owing to the liver occupying the exceptional position it does in relation to venous blood
that its special condition is attributable. What
essentially concerns us with reference to diabetes is,
that if oxygenated or imperfectly de-arterialised
blood pass to the liver through the portal vein, the
transformation of its amyloid substance into sugar
is occasioned, and glycosuria produced. Naturally
the liver receives thoroughly formed venous blood
through the portal vein. This is what is needed
for the proper exercise of its functional capacity,
and under this condition the formation of sugar is
not promoted.

We shall see later on how an imperfectly de-arterialised state of the blood contained in the portal
vein may be brought about to account for diabetes.
At present I am only dealing with the fact, that
blood in such a condition suffices to give rise to
glycosuria. I do not commit myself to any explanation of the precise *modus operandi*. The fact itself
is all that is actually wanted. That oxygenated
blood promotes the transformation of amyloid substance into sugar may be taken as established, but
whether as regards the sugar absorbed from the
alimentary canal it interferes with the formation of
the substance from it, or whether the formation
takes place and then the transformation back again
into sugar cannot be stated. To explain more fully
what I mean, let me represent the point as follows.
We know that under natural circumstances the
sugar derived from ingestion and absorbed from the
alimentary canal becomes converted into amyloid

substance. This I have previously spoken of as the first step of an assimilative action which is carried out by the liver. At present our knowledge does not enable us to follow the assimilative action on, and state what next occurs, but this, to my own mind, has been sufficiently established, that sugar gives origin to fat. In diabetes the assimilation of sugar fails to be carried out. It may be that the condition of the blood interferes with the accomplishment of the first step, or the formation of amyloid substance, and thus leads simply to the passage of the principle uninfluenced through the liver; or, it may be that conversion into amyloid substance occurs, and that this through the condition of the blood is brought back again into sugar. The fact stands that the sugar from ingestion is not stopped from reaching the general circulation as it ought to be, and I incline to the opinion that a simple passage through the liver is what occurs.

I may refer here to the production of glycosuria by an opposite condition to that which I have been commenting upon. In my former writings I have stated that glycosuria can be produced by obstructing the respiration to the extent of occasioning partial asphyxia. This may seem inconsistent with what I have been saying about oxygenated blood. No inconsistency, however, in reality exists, but apart from this, the statements are statements of fact, and if they appear irreconcilable the fault must lie with the power of explanation. Because the flow of oxygenated blood through the liver leads to the metamorphosis of the amyloid substance contained in the hepatic cells into sugar, and venous blood under natural conditions does not, it does not follow that amyloid substance is not convertible into sugar by the influence of direct contact with venous blood. Indeed, when amyloid substance is mixed with blood, no matter whether it be arterial or venous, its transformation into sugar is very speedily excited. Now, the effect of obstruction of the breathing is to produce impeded flow of blood through the lungs and a corresponding general venous congestion, and under this state the hepatic cells will be subjected to undue pressure, which may lead to more or less transudation and direct admixture of their contents with the blood. Thus the same effect may be induced as by the experimental injection of amyloid substance into the jugular vein, and this is known to be attended with the production of glycosuria.

We now come to the question of what bearing the fact relating to the production of glycosuria by the presence of oxygenated blood in the portal vein which I have been laying so much stress upon has in connection with human diabetes. Having reached this point we are brought into contact with the nervous system.

It has long been known that diabetes may be experimentally induced by puncturing a certain part of the floor of the fourth ventricle. The discovery of this fact by Bernard added greatly to his early renown, and strangely enough he was led up to it by a train of reasoning which afterwards proved to be groundless. For a time Bernard's experiment simply supplied an isolated piece of information, but later on, in endeavouring to find out the manner in which the puncture of the medulla oblongata acted, I discovered that injury of certain parts of the sympathetic system also produced glycosuria. I looked upon the medulla oblongata as a centre which must exercise an influence upon the liver, and directed my experiments to ascertain the channel through which this influence was exerted. In the course of my experiments I was led, from negative results obtained in other ways to try the effect of dividing the sympathetic filaments ascending from the superior thoracic ganglion to accompany the vertebral artery in its canal formed by the foramina in the transverse processes of the cervical vertebræ. This I found produced strongly marked glycosuria. I also, in the same course of experiments noticed that glycosuria followed removal of the superior cervical ganglion. From division of the gangliated cord in the chest, I likewise, upon some occasions, obtained a similar result, but in this group of experiments there was a want of the accord noticeable in the others. Division of all the nerves immediately belonging to the liver as they passed to the organ in company with the hepatic artery, hepatic duct, and portal vein in every instance failed to occasion glycosuria. My results were communicated to the Royal Society in 1859, and also published in the 'Guy's Hospital Reports' for that year. They have been confirmed by the subsequent experiments of Cyon and Aladoff, Schiff, and Eckhard, and I sometimes notice the experiments of these authorities referred to as though they furnished the original knowledge upon the subject.

I have said that glycosuria followed division of

the nerves passing from the superior thoracic ganglion to the vertebral artery.   On dividing the nerves
higher up in the neck—viz. within the vertebral
canals, which was effected by first carefully ligaturing the vertebral arteries below, and then with a
blunt hook tearing through all the structures passing
through a vertebral foramen, I did not observe
the production of saccharine urine unless the carotid
arteries were ligatured.   This was the issue of what
was noticed in several experiments, and I was
obliged simply to record it without being able satisfactorily to account for it.   I may further mention
that I was struck with the fact that in no instance
did I notice glycosuria when the operation was
performed above the foramen in the transverse process of the atlas.

Now, it has long been known that Bernard's
puncture induces a hyperæmic state of the chylopoietic viscera.   Schiff asserted, many years ago,
that lesion of the nervous centre in the region of
Bernard's puncture was accompanied with a dilatation of the small vessels of the intestine and liver,
producing a kind of paralytic hyperæmia of these
organs.   The same condition has been noticed to be
produced through the medium of injury of the sympathetic.   Schiff believed that from the hyperæmic
state he had noticed to exist a ferment was developed
which acted upon the amyloid substance of the liver,
exciting its transformation into sugar, and so accounting for the production of glycosuria.   This
view, which brings in the development of a ferment,
stands without support, and I will proceed to show

8

how the production of glycosuria may be otherwise explained.

One of the main points I have brought forward in these lectures is, that the effect of blood unduly charged with oxygen reaching the liver by the portal vein is to occasion glycosuria. It happens that this is just the state into which the portal blood is thrown by vaso-motor paralysis affecting the vessels of the chylopoietic viscera, and such, I consider, constitutes the key to the explanation of the saccharine condition of the urine in diabetes. It may be observed by superficial examination in the case of division of the sympathetic in the neck that not only is there a hyperæmic condition of the ear, but that the veins contain much redder blood than natural. In fact, the blood passes with such velocity and in such volume through the affected part that it does not become de-arterialised. A similar state existing in connection with the vessels of the chylopoietic viscera will give what is sufficient to produce glycosuria. Without any new agent being brought into the question the simple passage of blood through the vessels in such a manner as to cause it to arrive in the portal vein in an imperfectly de-arterialised condition will supply all that is wanted to account for the unnatural passage of sugar. In the vaso-motor paralysis, which, observation shows, is produced by lesions of the nervous system that give rise to glycosuria, we have a condition that leads to the presence of imperfectly de-arterialised blood in the portal vein, and in this presence of imperfectly de-arterialised blood in the portal vein we have a condition that suffices to determine the escape of sugar from the liver in a manner to produce a diabetic state of the urine. Physiologists have referred the production of glycosuria, under the circumstances alluded to, to hyperæmia of the liver. Doubtless, hyperæmia of the liver accompanies the exalted flow of blood noticed through the other viscera of the abdomen, but it is not specially this which is required to account for the phenomenon. I have already stated that I have not found glycosuria follow division of all the nerves passing in the lesser omentum to the liver—an operation which might be expected to cause hyperæmia of the organ by paralysis of the coats of its artery.

Such is the argument which leads us up to the consideration of glycosuria as a phenomenon of diabetes mellitus. I conceive it may be assumed that what is true for diabetes artificially induced may be taken as also holding good for diabetes of pathological occurrence. There is no reason to believe that the source of the glycosuria in the two instances is different, and it appears to me that support to this conclusion is given by the state in which the tongue is found in the most aggravated cases of the disease. Frequently, for instance, it may be noticed that the tongue presents an exceedingly injected appearance. Its bright red colour points to the existence of a hyperæmic state—the state that would be occasioned by a vaso-motor paralysis. The idea suggests itself from the appearance presented that the blood is flowing through the organ without being properly deprived of its arterial character. Now, just such a condition existing in

connection with the chylopoietic organs would
suffice, as I have furnished evidence to show, to
give rise to glycosuria. That the tongue, belonging
as it does to the digestive system, and the chylopoietic organs should be affected alike is nothing for
us to be unprepared for. It is in the worst kind of
case, as I have said, that the tongue presents the
appearance indicated, and the tongue being thus
implicated may be taken as simply expressing the
existence of a wider extent of vaso-motor paralysis
than where no such appearance is noticed.

The problem remaining to be solved is the nature
and seat of the lesion of the nervous system which
forms the primary morbid condition in diabetes.
There are various considerations connected with the
disease which point to the brain as being the most
likely part from which the morbid influence starts.
We know that by operating upon the medulla
oblongata and the sympathetic system diabetes may
be artificially induced, but it is not necessary that
attention should be confined to these structures, for
it is also known that in them only resides a part of
the vaso-motor system. The observations of Eulen-
berg, which are corroborated by those of Brown-
Sequard, have shown that the state of the arteries
is affected by lesion of certain parts of the grey
matter of the brain, and it is suggested that the
vaso-motor nerves distributed in the sympathetic,
besides being connected with the spinal cord and
medulla oblongata, pass up to spots at the surface
of the brain which stand in the position of cerebral
vaso-motor centres.

I certainly incline to the opinion that some kind

of textural change in the brain stands at the foundation of diabetes, and there are two ways in which
the effect may be produced. The vaso-motor system
exerts an influence upon the arteries which gives
them their tonus or keeps them in a certain state
of contraction. The effect of destruction of the
centres or tracts is to lead to arterial dilatation by
causing direct paralysis of the muscular coat, whilst
that of irritation is the converse. These results are
in harmony with what occurs under similar circumstances in connection with the cerebro-spinal system,
and are therefore intelligible enough to us. Arterial
dilatation, however, may be induced in another
way, viz. by an action of the cerebro spinal system
controlling or inhibiting the activity of the vaso-
motor system. Reference to what may be perceived
in connection with the salivary glands will supply
us with an illustration of the occurrence of this
mode of dilatation.

Whilst the glands are in a state of quiescence the
arteries are maintained in a comparatively contracted
condition by the influence of the vaso-motor nerves
they receive from the sympathetic. Now, common
observation shows us that a flow of saliva, which
means an antecedent dilatation of the arteries, may
be excited by a stimulus applied to the surface of
the tongue, and so influencing the extremities of
the lingual gustatory nerve, and even by an impression starting from the brain and originating in a
passing thought about food. The explanation
which physiological science gives of this phenomenon
is that a stimulus is transmitted from the cerebro-
spinal system which produces an inhibitory action

upon vaso-motor centres or nerves that relax the muscular coat of the arteries, and thus permits dilatation to occur. This may be considered to constitute an active form of vaso-dilatation in contradistinction to that arising from paralysis induced by operations upon the sympathetic which lead to a simple removal of vaso-motor influence.

There being these two modes of action by which vaso-dilatation may be brought about, it may happen that diabetes may arise either from a lesion affecting and involving a loss of power in vaso-motor centres, or a lesion in some part or other of the cerebro-spinal system leading to an inhibitory influence being exerted upon them.

Dr. Dickinson states that he has recognised certain vascular and perivascular changes in the brains of those who have died of diabetes. My colleagues, Drs. Frederick Taylor and Goodhart, however, in an article " On the Nervous System in Diabetes," published in the 'Guy's Hospital Reports' for 1877, have disputed the validity of his conclusions. The point in question is out of my own line of investigation, and I cannot therefore pretend to offer an opinion upon it; but this much I may say, that I hope in the interest of science Dr. Dickinson will continue and extend the inquiry he has begun, for I am sure it will be admitted that he possesses qualifications which specially fit him for such work. The ground I consider has been prepared by physiology, and we must now look for the fruits derivable from the assistance of pathology.

I may state, and I desire that the statement should stand only for what it may happen to be worth, that my clinical experience has led me to form the idea that diabetes presents an alliance as regards the character of its progress to locomotor ataxia and progressive muscular atrophy. We have in each of the three cases different manifestations of disordered nerve action to deal with ; but as in the diseases named, diabetes, certainly as it occurs amongst young subjects, is a truly progressive affection. For instance, when such a subject falls under observation during an early period of the disease, it will be found that by dietetic management alone the urine may be deprived of sugar and the health restored. Later on sugar re-appears notwithstanding the most rigorous dietetic management shall be persistently adopted. It may happen now that the administration of opium, morphia, or codeia, will again remove the sugar. Again, however, the urine becomes saccharine, and the saccharine character gradually augments. With this the patient notices that he loses ground in general health. He experiences more or less of a return of the old symptoms which troubled him at the outset before his eyes were open to the nature of his complaint. No impression now can be made upon the disease as formerly, and it appears to me we can only assume that the pathological condition which is at the bottom of it must have gradually advanced. The ordinary duration in these progressive cases, I think I may say, is about two years. Sometimes it is much shorter, sometimes longer; and as in locomotor ataxia the condition may advance to a certain point and then remain stationary for a shorter or longer while, the patient during this time being able

by strict attention to treatment, to keep himself in
a fair state of health. These remarks do not apply
to the disease in elderly persons; for here, if the
proper management be pursued, the tendency is
towards progress in a right direction, if not even
recovery, instead of towards an unfavorable issue.

This brings to a conclusion the remarks which
you, Mr. President, by the honour you conferred
upon me in appointing me to deliver these lectures,
have afforded me the opportunity of making upon a
subject which has always been one of much interest
to me. Whatever may prove to be the final issue of
investigation, the views I have expressed are those
which after a close attention to the subject for now
upwards of twenty years I have been led, rightly or
wrongly, to entertain.

INDEX.

9

PRINTED BY J. E. ADLARD, BARTHOLOMEW CLOSE

London, New Burlington Street.
October, 1878.

SELECTION

FROM

MESSRS J. & A. CHURCHILL'S

General Catalogue

COMPRISING

ALL RECENT WORKS PUBLISHED BY THEM

ON THE

ART AND SCIENCE

OF

MEDICINE

INDEX

THE PRACTICE OF SURGERY :
a Manual by THOMAS BRYANT, F.R.C.S., Surgeon to Guy's Hospital.
Third Edition, 2 vols., crown 8vo, with 672 Engravings.   [1878]

THE PRINCIPLES AND PRACTICE OF SURGERY,
by WILLIAM PIRRIE, F.R.S.E., Professor of Surgery in the University
of Aberdeen. Third Edition, 8vo, with 490 Engravings, 28s.   [1873]

A SYSTEM OF PRACTICAL SURGERY,
by Sir WILLIAM FERGUSSON, Bart., F.R.C.S., F.R.S. Fifth Edition,
8vo, with 463 Engravings, 21s.   [1870]

OPERATIVE SURGERY,
by C. F. MAUNDER, F.R.C.S., Surgeon to the London Hospital.
Second Edition, post 8vo, with 164 Engravings, 6s.   [1872]

BY THE SAME AUTHOR.

SURGERY OF THE ARTERIES :
Lettsomian Lectures for 1875, on Aneurisms, Wounds, Hæmorrhages,
&c. Post 8vo, with 18 Engravings, 5s.   [1875]

THE SURGEON'S VADE-MECUM,
a Manual of Modern Surgery, by ROBERT DRUITT. Eleventh Edition,
fcap. 8vo, with 369 Engravings, 14s.   [1878]

THE SCIENCE AND PRACTICE OF SURGERY :
a complete System and Textbook by F. J. GANT, F.R.C.S., Senior Surgeon to the Royal Free Hospital. 8vo, with 470 Engravings, 24s. [1871]

OUTLINES OF SURGERY AND SURGICAL PATHOLOGY,
including the Diagnosis and Treatment of Obscure and Urgent
Cases, and the Surgical Anatomy of some Important Structures and
Regions, by F. LE GROS CLARK, F.R.S., Consulting Surgeon to St.
Thomas's Hospital. Second Edition, Revised and Expanded by the
Author, assisted by W. W. WAGSTAFFE, F.R.C.S., Assistant-Surgeon
to St. Thomas's Hospital. 8vo, 10s. 6d.   [1872]

CLINICAL AND PATHOLOGICAL OBSERVATIONS IN INDIA,
by Sir J. FAYRER, K.C.S.I., M.D., F.R.C.P. Lond., F.R.S.E., Honorary
Physician to the Queen. 8vo, with Engravings, 20s.   [1873]

TREATMENT OF WOUNDS :
Clinical Lectures, by SAMPSON GAMGEE, F.R.S.E., Surgeon to the
Queen's Hospital, Birmingham. Crown 8vo, with Engravings, 5s. [1878]

BY THE SAME AUTHOR,

FRACTURES OF THE LIMBS
and their Treatment. 8vo, with Plates, 10s. 6d.   [1871]

THE FEMALE PELVIC ORGANS,
their Surgery, Surgical Pathology, and Surgical Anatomy, in a
Series of Coloured Plates taken from Nature : with Commentaries,
Notes, and Cases, by HENRY SAVAGE, M.D. Lond., F.R.C.S., Consulting
Officer of the Samaritan Free Hospital. Third Edition, 4to, £1 15s.
            [1875]

SURGICAL EMERGENCIES

together with the Emergencies attendant on Parturition and the Treatment of Poisoning: a Manual for the use of General Practitioners, by WILLIAM P. SWAIN, F.R.C.S., Surgeon to the Royal Albert Hospital, Devonport. Second Edition, post 8vo, with 104 Engravings, 6s. 6d. [1876]

TRANSFUSION OF HUMAN BLOOD:

with Table of 50 cases, by Dr. ROUSSEL, of Geneva. Translated by CLAUDE GUINNESS, B.A. With a Preface by SIR JAMES PAGET, Bart. Crown 8vo, 2s. 6d. [1877]

ILLUSTRATIONS OF CLINICAL SURGERY,

consisting of Coloured Plates, Photographs, Woodcuts, Diagrams, &c., illustrating Surgical Diseases, Symptoms and Accidents; also Operations and other methods of Treatment. By JONATHAN HUTCHINSON, F.R.C.S., Senior Surgeon to the London Hospital. In Quarterly Fasciculi, 6s. 6d. each. Fasciculi I to X bound, with Appendix and Index, £3 10s. [1876-8]

PRINCIPLES OF SURGICAL DIAGNOSIS

especially in Relation to Shock and Visceral Lesions, by F. LE GROS CLARK, F.R.C.S., Consulting Surgeon to St. Thomas's Hospital. 8vo, 10s. 6d. [1870]

MINOR SURGERY AND BANDAGING:

a Manual for the Use of House-Surgeons, Dressers, and Junior Practitioners, by CHRISTOPHER HEATH, F.R.C.S., Surgeon to University College Hospital, and Holme Professor of Surgery in University College. Fifth Edition, fcap 8vo, with 86 Engravings, 5s. 6d. [1875]

BY THE SAME AUTHOR,

INJURIES AND DISEASES OF THE JAWS:

JACKSONIAN PRIZE ESSAY. Second Edition, 8vo, with 164 Engravings, 12s. [1872]

BY THE SAME AUTHOR.

A COURSE OF OPERATIVE SURGERY:

with 20 Plates drawn from Nature by M. LÉVEILLÉ, and coloured by hand under his direction. Large 8vo. 40s. [1877]

HARE-LIP AND CLEFT PALATE,

by FRANCIS MASON, F.R.C.S., Surgeon and Lecturer on Anatomy at St. Thomas's Hospital. With 66 Engravings, 8vo, 6s. [1877]

BY THE SAME AUTHOR,

THE SURGERY OF THE FACE:

with 100 Engravings. 8vo, 7s. 6d. [1878]

FRACTURES OF THE LOWER END OF THE RADIUS,
Fractures of the Clavicle, and on the Reduction of the Recent Inward Dislocations of the Shoulder Joint. By ALEXANDER GORDON, M.D., Professor of Surgery in Queen's College, Belfast. With Engravings, 8vo, 5s. [1875]

DISEASES AND INJURIES OF THE EAR,
by W. B. DALBY, F.R.C.S., M.B., Aural Surgeon and Lecturer on Aural Surgery at St. George's Hospital. Crown 8vo, with 21 Engravings, 6s. 6d. [1873]

AURAL SURGERY;
A Practical Treatise, by H. MACNAUGHTON JONES, M.D., Surgeon to the Cork Ophthalmic and Aural Hospital. With 46 Engravings, crown 8vo, 5s. [1878.]

BY THE SAME AUTHOR,

ATLAS OF DISEASES OF THE MEMBRANA TYMPANI.
In Coloured Plates, containing 62 Figures, with Text, crown 4to, 21s. [1878]

THE EAR:
its Anatomy, Physiology, and Diseases. A Practical Treatise, by CHARLES H. BURNETT, A.M., M.D., Aural Surgeon to the Presbyterian Hospital, and Surgeon in Charge of the Infirmary for Diseases of the Ear, Philadelphia. With 87 Engravings, 8vo, 18s. [1877]

EAR AND THROAT DISEASES.
Essays by LLEWELLYN THOMAS, M.D., Surgeon to the Central London Throat and Ear Hospital. Post 8vo, 2s. 6d. [1878]

CLUBFOOT:
its Causes, Pathology, and Treatment: Jacksonian Prize Essay by WM. ADAMS, F.R.C.S., Surgeon to the Great Northern Hospital. Second Edition, 8vo, with 106 Engravings and 6 Lithographic Plates, 15s. [1873]

ORTHOPÆDIC SURGERY:
Lectures delivered at St. George's Hospital, by BERNARD E. BRODHURST, F.R.C.S., Surgeon to the Royal Orthopædic Hospital. Second Edition,8vo, with Engravings, 12s. 6d. [1876]

OPERATIVE SURGERY OF THE FOOT AND ANKLE,
by HENRY HANCOCK, F.R.C.S., Consulting Surgeon to Charing Cross Hospital. 8vo, with Engravings, 15s. [1873]

THE TREATMENT OF SURGICAL INFLAMMATIONS
by a New Method, which greatly shortens their Duration, by FURNEAUX JORDAN, F.R.C.S., Professor of Surgery in Queen's College, Birmingham. 8vo, with Plates, 7s. 6d. [1870]

BY THE SAME AUTHOR,

SURGICAL INQUIRIES.
With numerous Lithographic Plates. 8vo, 5s. [1873]

ORTHOPRAXY:
the Mechanical Treatment of Deformities, Debilities, and Deficiencies of the Human Frame, by H. HEATHER BIGG, Associate of the Institute of Civil Engineers. Third Edition, with 319 Engravings, 8vo, 15s. [1877]

ORTHOPÆDIC SURGERY:
and Diseases of the Joints. Lectures by LEWIS A. SAYRE, M.D.,
Professor of Orthopædic Surgery, Fractures and Dislocations, and
Clinical Surgery, in Bellevue Hospital Medical College, New York.
With 274 Wood Engravings, 8vo, 20s. [1876]

INTERNAL ANEURISM:
Its Successful Treatment by Consolidation of the Contents of the Sac.
By T. JOLIFFE TUFNELL, F.R.C.S.I., President of the Royal College
of Surgeons in Ireland. With Coloured Plates. Second Edition,
royal 8vo, 5s. [1875]

DISEASES OF THE RECTUM,
by THOMAS B. CURLING, F.R.S., Consulting Surgeon to the London
Hospital. Fourth Edition, Revised, 8vo, 7s. 6d. [1876]

BY THE SAME AUTHOR,
DISEASES OF THE TESTIS, SPERMATIC CORD, AND SCROTUM.
Third Edition, with Engravings, 8vo, 16s. [1878]

STRICTURE OF THE URETHRA
and Urinary Fistulæ; their Pathology and Treatment: Jacksonian
Prize Essay by Sir HENRY THOMPSON, F.R.C.S., Emeritus Professor
of Surgery to University College. Third Edition, 8vo, with Plates,
10s. [1869]

BY THE SAME AUTHOR,
PRACTICAL LITHOTOMY AND LITHOTRITY;
or, An Inquiry into the best Modes of removing Stone from the
Bladder. Second Edition, 8vo, with numerous Engravings. 10s. [1871]

ALSO,
DISEASES OF THE URINARY ORGANS:
(Clinical Lectures). Fourth Edition, 8vo, with 2 Plates and 59
Engravings, 12s. [1876]

ALSO,
DISEASES OF THE PROSTATE:
their Pathology and Treatment. Fourth Edition, 8vo, with numerous
Plates, 10s. [1873]

ALSO,
THE PREVENTIVE TREATMENT OF CALCULOUS DISEASE
and the Use of Solvent Remedies. Second Edition, fcap. 8vo, 2s. 6d.
[1876]

STRICTURE OF THE URETHRA,
and other Diseases of the Urinary Organs, by REGINALD HARRISON,
F.R.C.S., Surgeon to the Liverpool Royal Infirmary. With 10 plates.
8vo, 7s. 6d. [1878.]

LITHOTOMY AND EXTRACTION OF STONE
from the Bladder, Urethra, and Prostate of the Male, and from the
Bladder of the Female, by W. POULETT HARRIS, M.D., Surgeon-Major
H.M. Bengal Medical Service. With Engravings, 8vo, 10s. 6d. [1876]

THE SURGERY OF THE RECTUM:
Lettsomian Lectures by HENRY SMITH, F.R.C.S., Professor of Surgery in King's College, Surgeon to King's College Hospital. Fourth Edition, fcap. 8vo, 5s. [1876]

FISTULA, HÆMORRHOIDS, PAINFUL ULCER,
Stricture, Prolapsus, and other Diseases of the Rectum: their Diagnosis and Treatment, by WM. ALLINGHAM, F.R.C.S., Surgeon to St. Mark's Hospital for Fistula, &c. Second Edition, 8vo, 7s. [1872]

KIDNEY DISEASES, URINARY DEPOSITS,
and Calculous Disorders by LIONEL S. BEALE, M.B., F.R.S., F.R.C.P., Physician to King's College Hospital. Third Edition, 8vo, with 70 Plates, 25s. [1868]

DISEASES OF THE BLADDER,
Prostate Gland and Urethra, including a practical view of Urinary Diseases, Deposits and Calculi, by F. J. GANT, F.R.C.S., Senior Surgeon to the Royal Free Hospital. Fourth Edition, crown 8vo, with Engravings, 10s. 6d. [1876]

RENAL DISEASES:
a Clinical Guide to their Diagnosis and Treatment by W. R. BASHAM, M.D., F.R.C.P., late Senior Physician to the Westminster Hospital. Post 8vo, 7s. [1870]

BY THE SAME AUTHOR,
THE DIAGNOSIS OF DISEASES OF THE KIDNEYS,
with Aids thereto. 8vo, with 10 Plates, 5s. [1872]

THE REPRODUCTIVE ORGANS
in Childhood, Youth, Adult Age, and Advanced Life (Functions and Disorders of), considered in their Physiological, Social, and Moral Relations, by WILLIAM ACTON, M.R.C.S. Sixth Edition, 8vo, 12s. [1875]

URINARY AND REPRODUCTIVE ORGANS:
their Functional Diseases, by D. CAMPBELL BLACK, M.D., L.R.C.S. Edin. Second Edition. 8vo, 10s. 6d. [1875]

LECTURES ON SYPHILIS,
and on some forms of Local Disease, affecting principally the Organs of Generation, by HENRY LEE, F.R.C.S., Surgeon to St. George's Hospital. With Engravings, 8vo, 10s. [1875]

SYPHILITIC NERVOUS AFFECTIONS:
Their Clinical Aspects, by THOMAS BUZZARD, M.D., F.R.C.P. Lond., Physician to the National Hospital for Paralysis and Epilepsy. Post 8vo, 5s. [1874]

SYPHILIS:
Harveian Lectures, by J. R. LANE, F.R.C.S., Surgeon to, and Lecturer on Surgery at, St. Mary's Hospital; Consulting Surgeon to the Lock Hospital. Fcap. 8vo, 3s. 6d. [1878]

PATHOLOGY OF THE URINE,
including a Complete Guide to its Analysis, by J. L. W. THUDICHUM,
M.D., F.R.C.P. Second Edition, rewritten and enlarged, with Engravings, 8vo, 15s.   [1877]

GENITO-URINARY ORGANS, INCLUDING SYPHILIS:
A Practical Treatise on their Surgical Diseases, designed as a Manual
for Students and Practitioners, by W. H. VAN BUREN, M.D., Professor of the Principles of Surgery in Bellevue Hospital Medical College, New York, and E. L. KEYES, M.D., Professor of Dermatology in
Bellevue Hospital Medical College, New York. Royal 8vo, with 140
Engravings, 21s.   [1874]

HISTOLOGY AND HISTO-CHEMISTRY OF MAN:
A Treatise on the Elements of Composition and Structure of the
Human Body, by HEINRICH FREY, Professor of Medicine in Zurich.
Translated from the Fourth German Edition by ARTHUR E. J.
BARKER, Assistant-Surgeon to University College Hospital. And
Revised by the Author. 8vo, with 608 Engravings, 21s.   [1874]

HUMAN PHYSIOLOGY:
A Treatise designed for the Use of Students and Practitioners of
Medicine, by JOHN C. DALTON, M.D., Professor of Physiology and
Hygiene in the College of Physicians and Surgeons, New York. Sixth
Edition, royal 8vo, with 316 Engravings, 20s.   [1875]

HANDBOOK FOR THE PHYSIOLOGICAL LABORATORY,
by E. KLEIN, M.D., F.R.S., Assistant Professor in the Pathological Laboratory of the Brown Institution, London; J. BURDON-SANDERSON,
M.D., F.R.S., Professor of Practical Physiology in University College,
London; MICHAEL FOSTER, M.D., F.R.S., Prælector of Physiology
in Trinity College, Cambridge; and T. LAUDER BRUNTON, M.D.,
F.R.S., Lecturer on Materia Medica at St. Bartholomew's Hospital;
edited by J. BURDON-SANDERSON. 8vo, with 123 Plates, 24s.   [1873]

PRACTICAL HISTOLOGY:
By WILLIAM RUTHERFORD, M.D., Professor of the Institutes of
Medicine in the University of Edinburgh. Second Edition, with
63 Engravings. Crown 8vo (with additional leaves for notes), 6s.
                                                                      [1876]
THE MARRIAGE OF NEAR KIN,
Considered with respect to the Laws of Nations, Results of Experience,
and the Teachings of Biology, by ALFRED H. HUTH. 8vo, 14s.   [1875]

MANUAL OF ANTHROPOMETRY:
A Guide to the Measurement of the Human Body, containing an
Anthropometrical Chart and Register, a Systematic Table of Mea-
surements, &c. By CHARLES ROBERTS, F.R.C.S., late Assistant
Surgeon to the Victoria Hospital for Children. With numerous
Illustrations and Tables. 8vo, 6s. 6d.   [1878]

§

PRINCIPLES OF HUMAN PHYSIOLOGY,

by W. B. CARPENTER, C.B., M.D., F.R.S. Eighth Edition by HENRY POWER, M.B., F.R.C.S., Examiner in Natural Science, University of Oxford, and in Natural Science and Medicine, University of Cambridge. 8vo, with 3 Steel Plates and 371 Engravings, 31s. 6d. [1876]

STUDENTS' GUIDE TO HUMAN OSTEOLOGY,

By WILLIAM WARWICK WAGSTAFFE, F.R.C.S., Assistant-Surgeon and Lecturer on Anatomy, St. Thomas's Hospital. With 23 Plates and 66 Engravings. Fcap. 8vo, 10s. 6d. [1875]

LANDMARKS, MEDICAL AND SURGICAL,

By LUTHER HOLDEN, F.R.C.S., Member of the Court of Examiners of the Royal College of Surgeons. Second Edition, 8vo, 3s. 6d. [1877]

BY THE SAME AUTHOR.

HUMAN OSTEOLOGY:

Comprising a Description of the Bones, with Delineations of the Attachments of the Muscles, the General and Microscopical Structure of Bone, and its Development. Fifth Edition, with 61 Lithographic Plates and 89 Engravings. 8vo, 16s. [1878]

PATHOLOGICAL ANATOMY:

Lectures by SAMUEL WILKS, M.D., F.R.S., Physician to, and Lecturer on Medicine at, Guy's Hospital; and WALTER MOXON, M.D., F.R.C.P., Physician to, and Lecturer on Materia Medica at, Guy's Hospital. Second Edition, 8vo, with Plates, 18s. [1875]

PATHOLOGICAL ANATOMY:

A Manual by C. HANDFIELD JONES, M.B., F.R.S., Physician to St. Mary's Hospital, and EDWARD H. SIEVEKING, M.D., F.R.C.P., Physician to St. Mary's Hospital. Edited by J. F. PAYNE, M.D., F.R.C.P., Assistant Physician and Lecturer on General Pathology at St. Thomas's Hospital. Second Edition, crown 8vo, with 195 Engravings, 16s. [1875]

POST-MORTEM EXAMINATIONS:

a Description and Explanation of the Method of Performing them, with especial Reference to Medico-Legal Practice. By Professor RUDOLPH VIRCHOW, of Berlin. Fcap 8vo, 2s. 6d. [1876]

STUDENT'S GUIDE TO SURGICAL ANATOMY:

a Text-book for the Pass Examination, by E. BELLAMY, F.R.C.S., Surgeon and Lecturer on Anatomy at Charing Cross Hospital. Fcap 8vo, with 50 Engravings, 6s. 6d. [1873]

DIAGRAMS OF THE NERVES OF THE HUMAN BODY,

Exhibiting their Origin, Divisions, and Connexions, with their Distribution, by WILLIAM H. FLOWER, F.R.S., Conservator of Museum, Royal College of Surgeons. Second Edition, roy. 4to, 12s. [1872]

MEDICAL ANATOMY,

by FRANCIS SIBSON, M.D., F.R.C.P., F.R.S. Imp. folio, with 21 coloured Plates, cloth, 42s., half-morocco, 50s. [1869]

PRACTICAL ANATOMY:
a Manual of Dissections by CHRISTOPHER HEATH, F.R.C.S., Surgeon
to University College Hospital, and Holme Professor of Surgery in
University College.   Fourth Edition, crown 8vo, with 16 Coloured
Plates and 264 Engravings, 14s.   [1877]

AN ATLAS OF HUMAN ANATOMY:
illustrating most of the ordinary Dissections, and many not usually
practised by the Student.   To be completed in 12 or 13 Bi-monthly
Parts, each containing 4 Coloured Plates, with Explanatory Text. By
RICKMAN J. GODLEE, M.S., F.R.C.S., Assistant Surgeon to University
College Hospital, and Senior Demonstrator of Anatomy in University
College.   Imp. 4to, 7s. 6d. each Part.   [1877-8]

THE ANATOMIST'S VADE-MECUM:
a System of Human Anatomy by ERASMUS WILSON, F.R.C.S., F.R.S.
Ninth Edition, by G. BUCHANAN, M.A., M.D., Professor of Clinical
Surgery in the University of Glasgow, and HENRY E. CLARK, F.F.P.S.,
Lecturer on Anatomy at the Glasgow Royal Infirmary School of
Medicine.   Crown 8vo, with 371 Engravings, 14s.   [1873]

ATLAS OF TOPOGRAPHICAL ANATOMY,
after Plane Sections of Frozen Bodies.   By WILHELM BRAUNE,
Professor of Anatomy in the University of Leipzig.   Translated by
EDWARD BELLAMY, F.R.C.S., Surgeon to, and Lecturer on Anatomy,
&c., at, Charing Cross Hospital.   With 34 Photo-lithographic Plates
and 46 Woodcuts.   Large Imp. 8vo, 40s.   [1877]

THE ANATOMICAL REMEMBRANCER;
or, Complete Pocket Anatomist.   Eighth Edition, 32mo, 3s. 6d.   [1876]

THE STUDENT'S GUIDE TO THE PRACTICE OF MEDICINE,
by MATTHEW CHARTERIS, M.D., Professor of Medicine in Anderson's
College, and Lecturer on Clinical Medicine in the Royal Infirmary,
Glasgow. With Engravings on Copper and Wood, fcap. 8vo, 6s. 6d. [1877]

THE MICROSCOPE IN MEDICINE,
by LIONEL S. BEALE, M.B., F.R.S., Physician to King's College
Hospital.   Fourth Edition, with 86 Plates, 8vo, 21s.   [1877]

HOOPER'S PHYSICIAN'S VADE-MECUM;
or, Manual of the Principles and Practice of Physic, Ninth Edition
by W. A. GUY, M.B., F.R.S., and JOHN HARLEY, M.D., F.R.C.P.
Fcap 8vo, with Engravings, 12s. 6d.   [1874]

A NEW SYSTEM OF MEDICINE;
entitled Recognisant Medicine, or the State of the Sick, by
BHOLANOTH BOSE, M.D., Indian Medical Service.   8vo, 10s. 6d.   [1877]
                    BY THE SAME AUTHOR.

PRINCIPLES OF RATIONAL THERAPEUTICS.
Commenced as an Inquiry into the Relative Value of Quinine and
Arsenic in Ague.   8vo. 4s.   [1877]

THE STUDENT'S GUIDE TO MEDICAL DIAGNOSIS,
  by SAMUEL FENWICK, M.D., F.R.C.P., Physician to the London
  Hospital. Fourth Edition, fcap. 8vo, with 106 Engravings, 6s. 6d. [1876]

A MANUAL OF MEDICAL DIAGNOSIS,
  by A. W. BARCLAY, M.D., F.R.C.P., Physician to, and Lecturer on
  Medicine at, St. George's Hospital. Third Edition, fcap 8vo, 10s. 6d.
  [1876]

CLINICAL MEDICINE:
  Lectures and Essays by BALTHAZAR FOSTER, M.D., F.R.C.P. Lond.,
  Professor of Medicine in Queen's College, Birmingham. 8vo, 10s. 6d.
  [1874]

CLINICAL STUDIES:
  Illustrated by Cases observed in Hospital and Private Practice, by Sir
  J. ROSE CORMACK, M.D., F.R.S.E., Physician to the Hertford British
  Hospital of Paris. 2 vols., post 8vo, 20s.   [1876]

CLINICAL REMINISCENCES:
  By PEYTON BLAKISTON, M.D., F.R.S. Post 8vo, 3s. 6d.   [1878]

ROYLE'S MANUAL OF MATERIA MEDICA AND THERAPEUTICS.
  Sixth Edition by JOHN HARLEY, M.D., F.R.C.P., Assistant Physician
  to, and Joint Lecturer on Physiology at, St. Thomas's Hospital. Crown
  8vo, with 139 Engravings, 15s.   [1876]

PRACTICAL THERAPEUTICS:
  A Manual by E. J. WARING, M.D., F.R.C.P. Lond. Third Edition,
  fcap 8vo, 12s. 6d.   [1871]

THE ELEMENTS OF THERAPEUTICS.
  A Clinical Guide to the Action of Drugs, by C. BINZ, M.D., Professor
  of Pharmacology in the University of Bonn. Translated and Edited
  with Additions, in Conformity with the British and American Phar-
  macopœias, by EDWARD I. SPARKS, M.A., M.B. Oxon., formerly
  Radcliffe Travelling Fellow. Crown 8vo, 8s. 6d.   [1877]

THE STUDENT'S GUIDE TO MATERIA MEDICA,
  by JOHN C. THOROWGOOD, M.D., F.R.C.P. Lond., Physician to the
  City of London Hospital for Diseases of the Chest. Fcap 8vo, with
  Engravings, 6s. 6d.   [1874]

MATERIA MEDICA AND THERAPEUTICS:
  (Vegetable Kingdom), by CHARLES D. F. PHILLIPS, M.D., F.R.C.S.E.
  8vo, 15s.   [1874]

DENTAL MATERIA MEDICA AND THERAPEUTICS,
  Elements of, by JAMES STOCKEN, L.D.S.R.C.S., Lecturer on Dental
  Materia Medica and Therapeutics to the National Dental Hospital.
  Second Edition, Fcap 8vo, 6s. 6d.   [1876]

THE DISEASES OF CHILDREN:
  A Practical Manual, with a Formulary, by EDWARD ELLIS, M.D.,
  late Senior Physician to the Victoria Hospital for Children. Third
  Edition, crown 8vo, 7s. 6d.   [1878]

THE WASTING DISEASES OF CHILDREN,

by EUSTACE SMITH, M.D., F.R.C.P. Lond., Physician to the King of the Belgians, Physician to the East London Hospital for Children. Third Edition, post 8vo.   [In the Press.]

BY THE SAME AUTHOR,

CLINICAL STUDIES OF DISEASE IN CHILDREN.

Post 8vo, 7s. 6d.   [1876]

INFANT FEEDING AND ITS INFLUENCE ON LIFE;

or, the Causes and Prevention of Infant Mortality, by CHARLES H. F. ROUTH, M.D., Senior Physician to the Samaritan Hospital for Women and Children. Third Edition, fcap 8vo, 7s. 6d.   [1876]

COMPENDIUM OF CHILDREN'S DISEASES:

A Handbook for Practitioners and Students, by JOHANN STEINER, M.D., Professor in the University of Prague. Translated from the Second German Edition by LAWSON TAIT, F.R.C.S., Surgeon to the Birmingham Hospital for Women. 8vo, 12s. 6d.   [1874]

THE DISEASES OF CHILDREN:

Essays by WILLIAM HENRY DAY, M.D., Physician to the Samaritan Hospital for Diseases of Women and Children. Second Edition, fcap 8vo.
                                                               [In the Press.]

PUERPERAL DISEASES:

Clinical Lectures by FORDYCE BARKER, M.D., Obstetric Physician to Bellevue Hospital, New York. 8vo, 15s.   [1874]

THE STUDENT'S GUIDE TO THE PRACTICE OF MIDWIFERY,

by D. LLOYD ROBERTS, M.D., F.R.C.P., Physician to St. Mary's Hospital, Manchester. Second Edition, fcap. 8vo, with 95 Engravings.
                                                              [In the Press.]

OBSTETRIC MEDICINE AND SURGERY,

Their Principles and Practice, by F. H. RAMSBOTHAM, M.D., F.R.C.P. Fifth Edition, 8vo, with 120 Plates, 22s.   [1867]

OBSTETRIC SURGERY:

A Complete Handbook, giving Short Rules of Practice in every Emergency, from the Simplest to the most Formidable Operations connected with the Science of Obstetricy, by CHARLES CLAY, Ext.L.R.C.P. Lond., L.R.C.S.E., late Senior Surgeon and Lecturer on Midwifery, St. Mary's Hospital, Manchester. Fcap 8vo, with 91 Engravings, 6s. 6d.
                                                                      [1874]

SCHROEDER'S MANUAL OF MIDWIFERY,

including the Pathology of Pregnancy and the Puerperal State. Translated by CHARLES H. CARTER, B.A., M.D. 8vo, with Engravings, 12s. 6d.   [1873]

A HANDBOOK OF UTERINE THERAPEUTICS,

and of Diseases of Women, by E. J. TILT, M.D., M.R.C.P. Fourth Edition, post 8vo, 10s.   [1878]

BY THE SAME AUTHOR,

THE CHANGE OF LIFE

in Health and Disease: a Practical Treatise on the Nervous and other Affections incidental to Women at the Decline of Life. Third Edition, 8vo, 10s. 6d.   [1870]

OBSTETRIC OPERATIONS,
including the Treatment of Hæmorrhage, and forming a Guide to the Management of Difficult Labour; Lectures by ROBERT BARNES, M.D., F.R.C.P., Obstetric Physician and Lecturer on Obstetrics and the Diseases of Women and Children at St. George's Hospital. Third Edition, 8vo, with 124 Engravings, 18s. [1875]

BY THE SAME AUTHOR,
MEDICAL AND SURGICAL DISEASES OF WOMEN:
a Clinical History. Second Edition, 8vo, with 181 Engravings, 28s. [1878]

OBSTETRIC APHORISMS:
for the Use of Students commencing Midwifery Practice by J. G. SWAYNE, M.D., Consulting Physician-Accoucheur to the Bristol General Hospital, and Lecturer on Obstetric Medicine at the Bristol Medical School. Sixth Edition, fcap 8vo, with Engravings, 3s. 6d. [1876]

DISEASES OF THE OVARIES:
their Diagnosis and Treatment, by T. SPENCER WELLS, F.R.C.S., Surgeon to the Queen's Household and to the Samaritan Hospital. 8vo, with about 150 Engravings, 21s. [1872]

PRACTICAL GYNÆCOLOGY:
A Handbook of the Diseases of Women, by HEYWOOD SMITH, M.D. Oxon., Physician to the Hospital for Women and to the British Lying-in Hospital. With Engravings, crown 8vo, 5s. 6d. [1877]

RUPTURE OF THE FEMALE PERINEUM,
Its treatment, immediate and remote, by GEORGE G. BANTOCK, M.D., Surgeon (for In-patients) to the Samaritan Free Hospital for Women and Children. With 2 plates, 8vo, 3s. 6d. [1878.]

INFLUENCE OF POSTURE ON WOMEN
In Gynecic and Obstetric Practice, by J. H. AVELING, M.D., Physician to the Chelsea Hospital for Women, Vice-President of the Obstetrical Society of London. 8vo, 6s. [1878]

HANDBOOK FOR NURSES FOR THE SICK,
by ZEPHERINA P. VEITCH. Second Edition, crown 8vo, 3s. 6d. [1876]

A MANUAL FOR HOSPITAL NURSES
and others engaged in Attending on the Sick by EDWARD J. DOMVILLE, L.R.C.P., M.R.C.S., Surgeon to the Exeter Lying-in Charity. Third Edition, crown 8vo, 2s. 6d. [1879]

THE NURSE'S COMPANION:
A Manual of General and Monthly Nursing, by CHARLES J. CULLINGWORTH, Surgeon to St. Mary's Hospital, Manchester. Fcap. 8vo, 2s. 6d. [1876]

LECTURES ON NURSING,
by WILLIAM ROBERT SMITH, M.B., Honorary Medical Officer, Hospital for Sick Children, Sheffield. Second Edition, with 26 Engravings. Post 8vo, 6s. [1878]

A COMPENDIUM OF DOMESTIC MEDICINE
and Companion to the Medicine Chest; intended as a Source of Easy
Reference for Clergymen, and for Families residing at a Distance
from Professional Assistance, by JOHN SAVORY, M.S.A.   Ninth
Edition, 12mo, 5s.   [1878]

HOSPITAL MORTALITY
being a Statistical Investigation of the Returns of the Hospitals of
Great Britain and Ireland for fifteen years, by LAWSON TAIT, F.R.C.S.,
F.S.S. 8vo, 8s. 6d.   [1877]

THE COTTAGE HOSPITAL:
Its Origin, Progress, Management, and Work, by HENRY C. BURDETT,
the Seaman's Hospital, Greenwich. With Engravings, crown 8vo,
7s. 6d.   [1877]

WINTER COUGH:
(Catarrh, Bronchitis, Emphysema, Asthma), Lectures by HORACE
DOBELL, M.D., Consulting Physician to the Royal Hospital for Diseases
of the Chest. Third Edition, with Coloured Plates, 8vo, 1s. 6d.   [1875]

DISEASES OF THE CHEST:
Contributions to their Clinical History, Pathology, and Treatment, by
A. T. H. WATERS, M.D., F.R.C.P., Physician to the Liverpool Royal
Infirmary. Second Edition, 8vo, with Plates, 15s.   [1873]

CONSUMPTION:
Its Nature, Symptoms, Causes, Prevention, Curability, and Treatment.
By PETER GOWAN, M.D., B. Sc., late Physician and Surgeon in
Ordinary to the King of Siam. Crown 8vo. 5s.   [1878]

NOTES ON ASTHMA;
its Forms and Treatment, by JOHN C. THOROWGOOD, M.D. Lond.,
F.R.C.P., Physician to the Hospital for Diseases of the Chest, Victoria
Park. Third Edition, crown 8vo, 4s. 6d.   [1878]

ASTHMA
Its Pathology and Treatment, by J. B. BERKART, M.D., Assistant
Physician to the City of London Hospital for Diseases of the Chest.
8vo, 7s. 6d.   [1878]

PROGNOSIS IN CASES OF VALVULAR DISEASE OF THE
Heart, by THOMAS B. PEACOCK, M.D., F.R.C.P., Honorary Consulting Physician to St. Thomas's Hospital. 8vo, 3s. 6d.   [1877]

DISEASES OF THE HEART:
Their Pathology, Diagnosis, Prognosis, and Treatment (a Manual),
by ROBERT H. SEMPLE, M.D., F.R.C.P., Physician to the Hospital for
Diseases of the Throat. 8vo, 8s. 6d.   [1875]

CHRONIC DISEASE OF THE HEART:
Its Bearings upon Pregnancy, Parturition and Childbed. By ANGUS
MACDONALD, M.D., F.R.S.E., Physician to, and Clinical Lecturer on
the Diseases of Women at, the Edinburgh Royal Infirmary. With
Engravings, 8vo.   [1878]

PHTHISIS:
In a series of Clinical Studies, by AUSTIN FLINT, M.D., Professor of the Principles and Practice of Medicine and of Clinical Medicine in the Bellevue Hospital Medical College. 8vo, 16s. [1875]

BY THE SAME AUTHOR,

A MANUAL OF PERCUSSION AND AUSCULTATION,
of the Physical Diagnosis of Diseases of the Lungs and Heart, and of Thoracic Aneurism. Post 8vo, 6s. 6d. [1876]

GROWTHS IN THE LARYNX,
with Reports and an Analysis of 100 consecutive Cases treated since the Invention of the Laryngoscope by MORELL MACKENZIE, M.D. Lond., M.R.C.P., Physician to the Hospital for Diseases of the Throat. 8vo, with Coloured Plates, 12s. 6d. [1871]

DISEASES OF THE HEART AND AORTA,
By THOMAS HAYDEN, F.K.Q.C.P. Irel., Physician to the Mater Misericordiæ Hospital, Dublin. With 80 Engravings. 8vo, 25s. [1875]

DISEASES OF THE HEART
and of the Lungs in Connexion therewith—Notes and Observations by THOMAS SHAPTER, M.D., F.R.C.P. Lond., Senior Physician to the Devon and Exeter Hospital. 8vo, 7s. 6d. [1874]

DISEASES OF THE HEART AND AORTA:
Clinical Lectures by GEORGE W. BALFOUR, M.D., F.R.C.P., Physician to, and Lecturer on Clinical Medicine in, the Royal Infirmary, Edinburgh. 8vo, with Engravings, 12s. 6d. [1876]

PHYSICAL DIAGNOSIS OF DISEASES OF THE HEART.
Lectures by ARTHUR E. SANSOM, M.D., F.R.C.P., Assistant Physician to the London Hospital. Second Edition, with Engravings, fcap. 8vo, 4s. 6d. [1876]

TRACHEOTOMY,
especially in Relation to Diseases of the Larynx and Trachea, by PUGIN THORNTON, M.R.C.S., late Surgeon to the Hospital for Diseases of the Throat. With Photographic Plates and Woodcuts, 8vo, 5s. 6d. [1876]

SORE THROAT:
Its Nature, Varieties, and Treatment, including the Connexion between Affections of the Throat and other Diseases. By PROSSER JAMES, M.D., Lecturer on Materia Medica and Therapeutics at the London Hospital, Physician to the Hospital for Diseases of the Throat. Third Edition, with Coloured Plates, 5s. 6d. [1878.]

WINTER AND SPRING
on the Shores of the Mediterranean. By HENRY BENNET, M.D. Fifth Edition, post 8vo, with numerous Plates, Maps, and Engravings, 12s. 6d. [1874]

BY THE SAME AUTHOR,

TREATMENT OF PULMONARY CONSUMPTION
by Hygiene, Climate, and Medicine. Second Edition, 8vo, 5s. [1871]

PRINCIPAL HEALTH RESORTS
of Europe and Africa, and their Use in the Treatment of Chronic Diseases. A Handbook by THOMAS MORE MADDEN, M.D., M.R.I.A., Vice-President of the Dublin Obstetrical Society. 8vo, 10s. [1876]

THE BATH THERMAL WATERS:
Historical, Social, and Medical, by JOHN KENT SPENDER, M.D., Surgeon to the Mineral Water Hospital, Bath. With an Appendix on the Climate of Bath by the Rev. L. BLOMEFIELD, M.A., F.L.S., F.G.S. 8vo, 7s. 6d. [1877]

ENDEMIC DISEASES OF TROPICAL CLIMATES,
with their Treatment, by JOHN SULLIVAN, M.D., M.R.C.P. Post 8vo, 6s. [1877]

DISEASES OF TROPICAL CLIMATES
and their Treatment: with Hints for the Preservation of Health in the Tropics, by JAMES A. HORTON, M.D., Surgeon-Major, Army Medical Department. Post 8vo, 12s. 6d. [1874]

HEALTH IN INDIA FOR BRITISH WOMEN
and on the Prevention of Disease in Tropical Climates by EDWARD J. TILT, M.D., Consulting Physician-Accoucheur to the Farringdon General Dispensary. Fourth Edition, crown 8vo, 5s. [1875]

BURDWAN FEVER,
or the Epidemic Fever of Lower Bengal (Causes, Symptoms, and Treatment), by GOPAUL CHUNDER ROY, M.D., Surgeon Bengal Establishment. New Edition, 8vo, 5s. [1876]

BAZAAR MEDICINES OF INDIA
and Common Medical Plants: Remarks on their Uses, with Full Index of Diseases, indicating their Treatment by these and other Agents procurable throughout India, &c., by EDWARD J. WARING, M.D., F.R.C.P. Lond., Retired Surgeon H.M. Indian Army. Third Edition. Fcap 8vo, 5s. [1875]

SOME AFFECTIONS OF THE LIVER
and Intestinal Canal; with Remarks on Ague and its Sequelæ, Scurvy, Purpura, &c., by STEPHEN H. WARD, M.D. Lond., F.R.C.P., Physician to the Seamen's Hospital, Greenwich. 8vo, 7s. [1872]

DISEASES OF THE LIVER:
Lettsomian Lectures for 1872 by S. O. HABERSHON, M.D., F.R.C.P., Senior Physician to Guy's Hospital. Post 8vo, 3s. 6d. [1872]

BY THE SAME AUTHOR,

DISEASES OF THE STOMACH: DYSPEPSIA.
Second Edition, crown 8vo, 5s.

BY THE SAME AUTHOR,

PATHOLOGY OF THE PNEUMOGASTRIC NERVE,
being the Lumleian Lectures for 1876. Post 8vo, 3s. 6d. [1877]

BY THE SAME AUTHOR,

DISEASES OF THE ABDOMEN,
comprising those of the Stomach and other parts of the Alimentary Canal, Œsophagus, Cæcum, Intestines, and Peritoneum. Third Edition, with 5 Plates, 8vo, 21s. [1878]

FUNCTIONAL NERVOUS DISORDERS:
Studies by C. HANDFIELD JONES, M.B., F.R.C.P., F.R.S., Physician
to St. Mary's Hospital. Second Edition, 8vo, 18s.   [1870]

LECTURES ON DISEASES OF THE NERVOUS SYSTEM,
by SAMUEL WILKS, M.D., F.R.S., Physician to, and Lecturer on
Medicine at, Guy's Hospital. 8vo, 15s.   [1878]

NERVOUS DISEASES:
their Description and Treatment, by ALLEN MCLANE HAMILTON, M.D.,
Physician at the Epileptic and Paralytic Hospital, Blackwell's Island,
New York City. Roy. 8vo, with 53 Illustrations, 14s.   [1878]

NUTRITION IN HEALTH AND DISEASE:
A Contribution to Hygiene and to Clinical Medicine. By HENRY
BENNET, M.D. Third Edition. 8vo, 7s. Cheap Edition, Fcap. 8vo,
2s. 6d.   [1877]

FOOD AND DIETETICS,
Physiologically and Therapeutically Considered. By FREDERICK W.
PAVY, M.D., F.R.S., Physician to Guy's Hospital. Second Edition,
8vo, 15s.   [1875]

HEADACHES:
their Causes, Nature, and Treatment. By WILLIAM H. DAY, M.D.,
Physician to the Samaritan Free Hospital for Women and Children.
Second Edition, crown 8vo, with Engravings. 6s. 6d.   [1878]

IMPERFECT DIGESTION:
its Causes and Treatment by ARTHUR LEARED, M.D., F.R.C.P.,
Senior Physician to the Great Northern Hospital. Sixth Edition,
fcap 8vo, 4s. 6d.   [1875]

MEGRIM, SICK-HEADACHE,
and some Allied Disorders: a Contribution to the Pathology of Nerve-
Storms, by EDWARD LIVEING, M.D. Cantab., F.R.C.P., Hon. Fellow
of King's College, London. 8vo, with Coloured Plate, 15s.   [1873]

NEURALGIA AND KINDRED DISEASES
of the Nervous System: their Nature, Causes, and Treatment, with a
series of Cases, by JOHN CHAPMAN, M.D., M.R.C.P. 8vo, 14s.   [1873]

THE SYMPATHETIC SYSTEM OF NERVES,
and their Functions as a Physiological Basis for a Rational System of
Therapeutics by EDWARD MERYON, M.D., F.R.C.P., Physician to the
Hospital for Diseases of the Nervous System. 8vo, 3s. 6d.   [1872]

RHEUMATIC GOUT,
or Chronic Rheumatic Arthritis of all the Joints; a Treatise by
ROBERT ADAMS, M.D., M.R.I.A., late Surgeon to H.M. the Queen in
Ireland, and Regius Professor of Surgery in the University of Dublin.
Second Edition, 8vo, with Atlas of Plates, 21s.   [1872]

GOUT, RHEUMATISM,
and the Allied Affections; a Treatise by PETER HOOD, M.D. Crown
8vo, 10s. 6d.   [1871]

RHEUMATISM :
Notes by JULIUS POLLOCK, M.D., F.R.C.P., Senior Physician to, and
Lecturer on Medicine at, Charing Cross Hospital. Fcap. 8vo, 2s. 6d.
[1878.]
CANCER :
its varieties, their Histology and Diagnosis, by HENRY ARNOTT,
F.R.C.S., late Assistant-Surgeon to, and Lecturer on Morbid Anatomy
at, St. Thomas's Hospital.   8vo, with 5 Plates and 22 Engravings,
5s. 6d.   [1872]
CANCEROUS AND OTHER INTRA-THORACIC GROWTHS:
their Natural History and Diagnosis, by J. RISDON BENNETT, M.D.,
F.R.C.P., Member of the General Medical Council.   Post 8vo, with
Plates, 8s.   [1872]
CERTAIN FORMS OF CANCER,
with a New and successful Mode of Treating it, to which is prefixed a
Practical and Systematic Description of all the varieties of this Disease,
by ALEX. MARSDEN, M.D., F.R.C.S.E., Consulting Surgeon to the
Royal Free Hospital, and Senior Surgeon to the Cancer Hospital.
Second Edition, with Coloured Plates, 8vo, 8s. 6d.   [1873]
ATLAS OF SKIN DISEASES :
a series of Illustrations, with Descriptive Text and Notes upon Treat-
ment. By TILBURY FOX, M.D., F.R.C.P., Physician to the Department
for Skin Diseases in University College Hospital.   With 72 Coloured
Plates, royal 4to, half morocco, £6 6s.   [1877]
DISEASES OF THE SKIN :
a System of Cutaneous Medicine by ERASMUS WILSON, F.R.C.S.,
F.R.S. Sixth Edition, 8vo, 18s., with Coloured Plates, 36s.   [1867]
BY THE SAME AUTHOR,
LECTURES ON EKZEMA
and Ekzematous Affections : with an Introduction on the General
Pathology of the Skin, and an Appendix of Essays and Cases. 8vo,
10s. 6d.   [1870]
ALSO,
LECTURES ON DERMATOLOGY :   ·
delivered at the Royal College of Surgeons, 1870, 6s. ; 1871-3, 10s. 6d.,
1874-5, 10s. 6d. ; 1876-8, 10s. 6d.
ECZEMA :
by McCALL ANDERSON, M.D., Professor of Clinical Medicine in the
University of Glasgow.   Third Edition, 8vo, with Engravings, 7s. 6d.
[1874]
BY THE SAME AUTHOR,
PARASITIC AFFECTIONS OF THE SKIN
Second Edition, 8vo, with Engravings, 7s. 6d.   [1868]
PSORIASIS OR LEPRA,
by GEORGE GASKOIN, M.R.C.S., Surgeon to the British Hospital for
Diseases of the Skin.   8vo, 5s.   [1875]

MYCETOMA;

or, the Fungus Disease of India, by H. VANDYKE CARTER, M.D., Surgeon-Major H.M. Indian Army.   4to, with 11 Coloured Plates, 42s.
[1874]

CERTAIN ENDEMIC SKIN AND OTHER DISEASES

of India and Hot Climates generally, by TILBURY FOX, M.D., F.R.C.P.,
and T. FARQUHAR, M.D. (Published under the sanction of the Secre-
tary of State for India in Council).   8vo, 10s. 6d.   [1876]

DISEASES OF THE SKIN,

in Twenty-four Letters on the Principles and Practice of Cutaneous
Medicine, by HENRY EVANS CAUTY, M.R.C.S., Surgeon to the Liverpool Dispensary for Diseases of the Skin, 8vo, 12s. 6d.   [1874]

THE HAIR IN HEALTH AND DISEASE,

by E. WYNDHAM COTTLE, F.R.C.S., Senior Assistant Surgeon to the
Hospital for Diseases of the Skin, Blackfriars. Fcap. 8vo, 2s. 6d. [1877]

WORMS:

a Series of Lectures delivered at the Middlesex Hospital on Practical
Helminthology by T. SPENCER COBBOLD, M.D., F.R.S.   Post 8vo,
5s.   [1872]

THE LAWS AFFECTING MEDICAL MEN:

a Manual by ROBERT G. GLENN, LL.B., Barrister-at-Law; with a
Chapter on Medical Etiquette by Dr. A. CARPENTER.   8vo, 14s. [1871]

MEDICAL JURISPRUDENCE,

Its Principles and Practice, by ALFRED S. TAYLOR, M.D., F.R.C.P.,
F.R.S.   Second Edition, 2 vols., 8vo, with 189 Engravings, £1 11s. 6d.
[1873]

BY THE SAME AUTHOR,

A MANUAL OF MEDICAL JURISPRUDENCE.

Ninth Edition.   Crown 8vo, with Engravings, 14s.   [1874]

ALSO,

POISONS,

in Relation to Medical Jurisprudence and Medicine.   Third Edition,
crown 8vo, with 104 Engravings, 16s.   [1875]

MEDICAL JURISPRUDENCE :

Lectures by FRANCIS OGSTON, M.D., Professor of Medical Jurisprudence and Medical Logic in the University of Aberdeen.   Edited
by FRANCIS OGSTON, Jun., M.D., Assistant to the Professor of
Medical Jurisprudence and Lecturer on Practical Toxicology in the
University of Aberdeen.   8vo, with 12 Copper Plates, 18s.   [1878]

A TOXICOLOGICAL CHART,

exhibiting at one View the Symptoms, Treatment, and mode of
Detecting the various Poisons—Mineral, Vegetable, and Animal :
with Concise Directions for the Treatment of Suspended Animation,
by WILLIAM STOWE, M.R.C.S.E.   Thirteenth Edition, 2s.; on
roller, 5s.   [1872]

A HANDY-BOOK OF FORENSIC MEDICINE AND TOXICOLOGY,
by W. BATHURST WOODMAN, M.D., F.R.C.P., Assistant Physician
and Co-Lecturer on Physiology and Histology at the London Hospital;
and C. MEYMOTT TIDY, M.D., F.C.S., Professor of Chemistry and of
Medical Jurisprudence and Public Health at the London Hospital.
With 8 Lithographic Plates and 116 Engravings, 8vo, 31s. 6d.   [1877]

THE MEDICAL ADVISER IN LIFE ASSURANCE,
by EDWARD HENRY SIEVEKING, M.D., F.R.C.P., Physician to St.
Mary's and the Lock Hospitals; Physician-Extraordinary to the
Queen; Physician-in-Ordinary to the Prince of Wales, &c. Crown
8vo, 6s.   [1874]

IDIOCY AND IMBECILITY,
by WILLIAM W. IRELAND, M.D., Medical Superintendent of the
Scottish National Institution for the Education of Imbecile Children
at Larbert, Stirlingshire. With Engravings, 8vo, 14s.   [1877]

PSYCHOLOGICAL MEDICINE:
a Manual, containing the Lunacy Laws, the Nosology, Ætiology,
Statistics, Description, Diagnosis, Pathology (including Morbid Histology), and Treatment of Insanity, by J. C. BUCKNILL, M.D.,
F.R.S., and D. H. TUKE, M.D., F.R.C.P. Third Edition, 8vo, with
10 Plates and 34 Engravings, 25s.   [1873]

MADNESS:
in its Medical, Legal, and Social Aspects, Lectures by EDGAR
SHEPPARD, M.D., M.R.C.P., Professor of Psychological Medicine in
King's College; one of the Medical Superintendents of the Colney
Hatch Lunatic Asylum. 8vo, 6s. 6d.   [1873]

HANDBOOK OF LAW AND LUNACY;
or, the Medical Practitioner's Complete Guide in all Matters relating
to Lunacy Practice, by J. T. SABBEN, M.D., and J. H. BALFOUR
BROWNE, Barrister-at-Law. 8vo, 5s.   [1872]

INFLUENCE OF THE MIND UPON THE BODY
in Health and Disease, Illustrations designed to elucidate the Action
of the Imagination, by DANIEL HACK TUKE, M.D., F.R.C.P.
8vo, 14s.   [1872]

A MANUAL OF PRACTICAL HYGIENE,
by E. A. PARKES, M.D., F.R.S. Fifth Edition, by F. DE CHAUMONT,
M.D., Professor of Military Hygiene in the Army Medical School.
8vo, with 9 Plates and 112 Engravings, 18s.   [1878]

A HANDBOOK OF HYGIENE AND SANITARY SCIENCE,
by GEORGE WILSON, M.A., M.D., Medical Officer of Health for Mid-
Warwickshire. Third Edition, post 8vo, with Engravings, 10s. 6d.   [1877]

MICROSCOPICAL EXAMINATION OF DRINKING WATER:
A Guide, by JOHN D. MACDONALD, M.D., F.R.S., Assistant Professor of Naval Hygiene, Army Medical School. 8vo, with 24 Plates,
7s. 6d.   [1875]

HANDBOOK OF MEDICAL AND SURGICAL ELECTRICITY,

by HERBERT TIBBITS, M.D., F.R.C.P.E., Medical Superintendent of the National Hospital for the Paralysed and Epileptic. Second Edition 8vo, with 95 Engravings, 9s. [1877]

BY THE SAME AUTHOR.

A MAP OF ZIEMSSEN'S MOTOR POINTS OF THE HUMAN BODY:

a Guide to Localised Electrisation. Mounted on Rollers, 35 × 21. With 20 Illustrations, 5s. [1877]

CLINICAL USES OF ELECTRICITY;

Lectures delivered at University College Hospital by J. RUSSELL REYNOLDS, M.D. Lond., F.R.C.P., F.R.S., Professor of Medicine in University College. Second Edition, post 8vo, 3s. 6d. [1873]

MEDICO-ELECTRIC APPARATUS:

A Practical Description of every Form in Modern Use, with Plain Directions for Mounting, Charging, and Working, by SALT & SON, Birmingham. Second Edition, revised and enlarged, with 33 Engravings, 8vo, 2s. 6d. [1877]

A DICTIONARY OF MEDICAL SCIENCE;

containing a concise explanation of the various subjects and terms of Medicine, &c.; Notices of Climate and Mineral Waters; Formulæ for Officinal, Empirical, and Dietetic Preparations; with the Accentuation and Etymology of the terms and the French and other Synonyms, by ROBLEY DUNGLISON, M.D., LL.D. New Edition, royal 8vo, 28s. [1874]

A MEDICAL VOCABULARY;

being an Explanation of all Terms and Phrases used in the various Departments of Medical Science and Practice, giving their derivation, meaning, application, and pronunciation, by ROBERT G. MAYNE, M.D., LL.D. Fourth Edition, fcap 8vo, 10s. [1875]

ATLAS OF OPHTHALMOSCOPY,

by R. LIEBREICH, Ophthalmic Surgeon to St. Thomas's Hospital. Translated into English by H. ROSBOROUGH SWANZY, M.B. Dub. Second Edition, containing 59 Figures, 4to, £1 10s. [1870]

DISEASES OF THE EYE:

a Manual by C. MACNAMARA, F.R.C.S., Surgeon to Westminster Hospital. Third Edition, fcap. 8vo, with Coloured Plates and Engravings, 12s. 6d. [1876]

DISEASES OF THE EYE:

A Practical Treatise by HAYNES WALTON, F.R.C.S., Surgeon to St. Mary's Hospital and in charge of its Ophthalmological Department. Third Edition, 8vo, with 3 Plates and nearly 300 Engravings, 25s.
[1875]

HINTS ON OPHTHALMIC OUT-PATIENT PRACTICE,

by CHARLES HIGGENS, F.R.C.S., Ophthalmic Assistant Surgeon to, and Lecturer on Ophthalmology at, Guy's Hospital. 87 pp., fcap. 8vo, 2s. 6d. [1877]

OPHTHALMIC MEDICINE AND SURGERY:
a Manual by T. WHARTON JONES, F.R.C.S., F.R.S., Professor of Ophthalmic Medicine and Surgery in University College. Third Edition,
fcap. 8vo, with 9 Coloured Plates and 173 Engravings, 12s. 6d.   [1865]

DISEASES OF THE EYE:
A Treatise by J. SOELBERG WELLS, F.R.C.S., Ophthalmic Surgeon to
King's College Hospital and Surgeon to the Royal London Ophthalmic
Hospital. Third Edition, 8vo, with Coloured Plates and Engravings,
25s.   [1873]

BY THE SAME AUTHOR,

LONG, SHORT, AND WEAK SIGHT,
and their Treatment by the Scientific use of Spectacles. Fourth
Edition, 8vo, 6s.   [1873]

A SYSTEM OF DENTAL SURGERY,
by JOHN TOMES, F.R.S., and CHARLES S. TOMES, M.A., F.R.S., Lecturer on Dental Anatomy and Physiology at the Dental Hospital of
London. Second Edition, fcap 8vo, with 268 Engravings, 14s.   [1873]

DENTAL ANATOMY, HUMAN AND COMPARATIVE:
A Manual, by CHARLES S. TOMES, M.A., F.R.S., Lecturer on Dental
Anatomy and Physiology at the Dental Hospital of London. With
179 Engravings, crown 8vo, 10s. 6d.   [1876]

A MANUAL OF DENTAL MECHANICS,
with an Account of the Materials and Appliances used in Mechanical
Dentistry, by OAKLEY COLES, L.D.S., R.C.S., Surgeon-Dentist to
the Hospital for Diseases of the Throat. Second Edition, crown 8vo,
with 140 Engravings, 7s. 6d.   [1876]

HANDBOOK OF DENTAL ANATOMY
and Surgery for the use of Students and Practitioners by JOHN
SMITH, M.D., F.R.S. Edin., Surgeon-Dentist to the Queen in Scotland.
Second Edition, fcap 8vo, 4s. 6d.   [1871]

STUDENT'S GUIDE TO DENTAL ANATOMY AND SURGERY,
by HENRY SEWILL, M.R.C.S., L.D.S., Dentist to the West London
Hospital. With 77 Engravings, fcap. 8vo, 5s. 6d.   [1876]

OPERATIVE DENTISTRY:
A Practical Treatise, by JONATHAN TAFT, D.D.S., Professor of Operative Dentistry in the Ohio College of Dental Surgery. Third Edition,
thoroughly revised, with many additions, and 134 Engravings, 8vo,
18s.   [1877]

DENTAL CARIES
and its Causes: an Investigation into the influence of Fungi in the
Destruction of the Teeth, by Drs. LEBER and ROTTENSTEIN. Translated by H. CHANDLER, D.M.D., Professor in the Dental School of
Harvard University. With Illustrations, royal 8vo, 5s.   [1878.]

EPIDEMIOLOGY;
or, the Remote Cause of Epidemic Diseases in the Animal and in the
Vegetable Creation, by JOHN PARKIN, M.D., F.R.C.P.E. Part I,
Contagion—Modern Theories—Cholera—Epizootics. 8vo, 5s.   [1873]

The following CATALOGUES issued by Messrs CHURCHILL will be forwarded post free on application :

1. *Messrs Churchill's General List of nearly* 600 *works on Medicine, Surgery, Midwifery, Materia Medica, Hygiene, Anatomy, Physiology, Chemistry, &c., &c., with a complete Index to their Titles, for easy reference.*

N.B.—*This List includes Nos.* 2 *and* 3.

2. *Selection from Messrs Churchill's General List, comprising all recent Works published by them on the Art and Science of Medicine.*

3. *A descriptive List of Messrs Churchill's Works on Chemistry, Pharmacy, Botany, Photography, Zoology, and other branches of Science.*

4. *Messrs Churchill's Red-Letter List, giving the Titles of forthcoming New Works and New Editions.*

[Published every October.]

5. *The Medical Intelligencer, an Annual List of New Works and New Editions published by Messrs J. & A. Churchill, together with Particulars of the Periodicals issued from their House.*

[Sent in January of each year to every Medical Practitioner in the United Kingdom whose name and address can be ascertained. A large number are also sent to the United States of America, Continental Europe, India, and the Colonies.]

MESSRS CHURCHILL have a special arrangement with MESSRS LINDSAY & BLAKISTON, OF PHILADELPHIA, in accordance with which that Firm act as their Agents for the United States of America, either keeping in Stock most of Messrs CHURCHILL's Books, or reprinting them on Terms advantageous to Authors. Many of the Works in this Catalogue may therefore be easily obtained in America.

PRINTED BY J. E. ADLARD, BARTHOLOMEW CLOSE.